W9-AEC-267

THIS SIGNED FIRST EDITION OF

POPULATION WARS

A New Perspective on Competition and Coexistence

by

GREG GRAFFIN

HAS BEEN SPECIALLY BOUND AND PRODUCED BY THE PUBLISHER

POPULATION WARS

ALSO BY GREG GRAFFIN

Anarchy Evolution

POPULATION WARS

A New Perspective on Competition and Coexistence

GREG GRAFFIN

Thomas Dunne Books
St. Martin's Press
New York

THOMAS DUNNE BOOKS.
An imprint of St. Martin's Press.

www.thomasdunnebooks.com
www.stmartins.com

Designed by Omar Chapa

The Library of Congress Cataloging-in-Publication Data is available upon request.

ISBN: 978-1-250-01762-8 (hardcover)
ISBN: 978-1-250-01761-1 (e-book)

First Edition: September 2015

10 9 8 7 6 5 4 3 2 1

To Allison

CONTENTS

POPULATION

Any group of organisms coexisting at the same time and in the same place and capable of interbreeding with one another.

—David Sadava et al., *Life, the Science of Biology,* 8th ed.

A species in time and space is composed of numerous local populations, each one intercommunicating and intergrading with the others.

—Ernst Mayr, *Animal Species and Evolution*

Any population tends to grow according to the compound interest law as long as the individuals continue to live under the same conditions. . . . An increasing population cannot possibly, however, continue to live under the same conditions for many generations.

—Sewall Wright, *Evolution and the Genetics of Populations*

The so-called unit of evolution.

—Ernst Mayr, *What Evolution Is*

WAR

Any active hostility between living beings; a campaign against something pervasive and undesireable.

—*Concise Oxford English Dictionary*, 12th ed.

An act of violence intended to compel our opponent to fulfill our will.

—Carl von Clausewitz, 1827; *On War,*
trans. J. J. Graham

POPULATION WARS

INTRODUCTION

FINDING THE ENEMY

Living things have something profound in common: They are all members of populations. In the past these groups were easy to tell apart. We humans considered some of them evil, some friendly, some wild and untamed, and some put here simply to serve our own selfish needs. It was easy to do so because we were ignorant. We didn't have DNA tests to determine relatedness. We didn't have electron microscopes to view microbes living inside us. We didn't have radioactive isotopes to calibrate the time lines of migrations. And before about 1859, we didn't have a narrative that linked all organisms as related by descent from a common ancestor.

Today, as more facts of history and biology have come to light, it's much harder to draw a distinction between ourselves and those we consider "others." This has created an intellectual crisis on many levels. How, for instance, do we make sense of the data that show humans are healthier when they have symbiotic bacteria living all over and inside them? This implies that our "selves" are actually shared communities of different species. What do we make of the data that show nearly half of our genome is from viruses? We carry around in each of our cells a habitat that allows viral replication as a necessary mechanism in the perpetuation of our own species. Another portion of the human genome is from Neanderthals, long-dead ancestors who could rightly claim us all as carrying on a portion of their genetic heritage. How can we reconcile this growing list of facts that testify to the ubiquity of symbiosis with the old narratives of competition and the "struggle for existence"? I think humans—particularly those who consider themselves paragons of an intellectually advanced, modern society—are having a hard time with it. Rapidly accumulating biological data are causing a dissonance of sorts between old narratives, based on ignorance, that run deep in the collective consciousness of all societies, and the new implications coming from recent studies. Scientists in general do a great job of collecting empirical data, but generally fall short in explaining the implications of their findings. Because of this, ironically, the delusions of modern Western society grow stronger in the face of increased knowledge. This is particularly evident in the narrative of war.

Why do we go to war? There are many answers, but part of this book concerns one of them: because war is an inevitable property of humankind, an inheritance from our distant ancestors, and as such it's part of the interconnectedness of the biosphere throughout its long history. In other words, war is part of the symbiotic heritage of all life. Therefore we must look to coexisting populations and their interactions, historical as well as recent, human as well as other species', if we ever want a serious answer to the question above. Behaviors akin to warfare are found in species across the entire spectrum of the animal kingdom, so it's no surprise that humans exhibit them too.

Despite the ubiquity (and tacit inevitability) of population wars, a closer examination of some distantly related species reveals that there is as much interdependence in the biosphere as there is violence. The present is full of assimilations from populations of the past, and when we recognize some of these, it becomes apparent that there is hope for a less violent future for humankind. Hence the inevitability of population wars doesn't mean that our future has to be violently catastrophic. In order to achieve such a result, however, humans have to come to terms with some basic facts of population biology, and we need to see a shift in consciousness away from some of our most deeply rooted prejudices and bad habits that have come from the rapid expansion of our species.

Today we exist as a globally distributed species with a particular nasty propensity: When we can't see our enemies, we invent them. This illusory act of human nature allows us to justify attempts at eradicating, eliminating, or vanquishing other people or species. But such actions nearly always fail. A further goal of this book is to highlight those failures as a reason to take a fresh approach to one of the time-honored problems of human existence: defining "us" as distinct from "them." This is ultimately a question of biology. How distinct are different groups of living things? I hope to show you that the answer is, A lot less than you previously thought. If I'm successful, you'll see why we need to rethink the entire justification for war, not only the human military kind but also the thought of Darwin's "war of nature" or "struggle for existence," because war follows logically only from a notion of distinctness. If lines of distinction are blurred, whom (or what) are we fighting? In my view all types of conflict have to be recast in the light of coexistence and historical contingency. That is the message of this book.

We Americans love the war metaphor. As I write this, it seems that everyone is debating over how to deal with sinister forces that are attacking us on many fronts. We have the war on poverty, the war on drugs, the war on Ebola, the war on terror. Then there's the war on the middle class, the war on workers, the culture wars, the war on women, the war on our kids, the war on family, among many others. Given all these deadly conflicts we're engaged in, it's no wonder that we're considered a war-weary nation.[1] We turn to narratives of "good vs. evil," such as the Star Wars epics or the slew of Marvel comic-book movies, with almost religious fervor. These simplistic tales are emotionally satisfying to the general public, perhaps because they hark back to a time when it was easier to believe that the "bad guys" could be easily eradicated. Unfortunately, today we know better.

Are the enemies we face in these wars conquerable? The answer is no, but it seems that most people don't know it. These trite phrases—"the war on . . ."—make for attention-grabbing headlines because of their emotional appeal, not their intellectual accuracy. Most people assume that there is an easy victory to every war. Whether it's exterminating a pest from your house, eliminating a military force with a drone strike, or erasing an entire enemy population by dropping a nuclear warhead.

It should come as no surprise that none of those solutions have worked in the past, and they won't work in the future either. Although in the short term an enemy might appear to be vanquished, in the long term that same population rebounds and—even if changed, perhaps—it eventually rebuilds its numbers. What we commonly assume to be wars are an elemental part of life, and the ebb and flow of population size is due to these conflicts. In fact they typify an ongoing biological drama that has been occurring since the origin of the biosphere, nearly 4 billion years ago. Populations have an inherent tendency to expand, and eventually, overcoming the constraints of their environment, they come in contact with one another. History is written and calibrated by these constant population wars that have occurred and that are still occurring today. Our only hope of mitigating entrenched misunderstandings, and in due course eliminating human wars—the actual military type that has caused so much needless suffering—is first to understand more about the natural world and the population interactions that typify other species.

Debates rage in our country over whether or not to ban all flights from Ebola-plagued countries in Africa, as if that alone could stop the "enemy" microbe from spreading. The idea is simple: Equate an enemy population of microbes with an enemy host population and vanquish both at once. As of this writing, more than 10,600 people have died from that disease in Liberia, Guinea, and Sierra Leone. Four people, on different occasions, have brought it back to the United States, and one of them died. But in the process the dead man infected two nurses in Texas, one of whom flew on a domestic airplane to Ohio, and may have passed the Ebola virus on to others. How this story plays out is anyone's guess, but clearly we are witnessing a population war that is complicated. A microbe is increasing its species' range. An ever-increasing panmictic[2] human population is continuing to assimilate. And the American public—the vast majority of whom have a very low probability of ever coming in contact with an Ebola victim—are frightened beyond belief, and busy themselves with judgments about who is the enemy. Blame goes to the government—the CDC should have done more to stop the spread. Blame goes to the individual—the victim, now dead, should never have carried the disease to our country.[3] Blame goes to the doctors—why did the Texas hospital staff allow a man infected with Ebola to leave the hospital instead of immediately putting him in quarantine? In fact there is enough blame to dole out to multiple enemies in order to justify eliminating all of them in this war on Ebola. But that won't make the disease go away. Blame doesn't solve population wars; it simply helps preserve—and foster—the illusion of an enemy.

Debates rage about the culprits in other diseases as well. Who will play the role of the victim and who the perpetrator in the war on AIDS or the war on tuberculosis? Is the flu shot an effective way to vanquish a persistent enemy, which attacks us every fall and winter? Are chronic intestinal ailments and digestive dysfunctions caused by the food industries? Or can those wars be won simply by eating more yogurt probiotics?

There is much to be considered with respect to human and microbe interactions. But I don't want to give the impression that this book deals only with diseases, for there are many other types of population wars. Debates rage over the need for protection of endangered species. Livestock owners blame their lack of profits on the wolf, or the mountain lion, both majestic animals that live by eating other mammals, including domestic cattle and sheep. Protection of predatory species, and in some cases reintroduction of them into their former geographic range, now occupied by ranches, draws ire from property owners and applause from environmentalists (many of whom live far from the regions in question). Again we see a multidimensional problem that isn't simple to convey in a catchy one-liner. The biological dimension of this population war—predators and their prey—has been going on since the origin of vertebrate animals at least 540 million years ago. The human dimension of this war is framed in terms of environmental health and restoration practices vs. the livelihood of farmers and ranchers and how public land should be used. Any change to one of these terms has a profound effect on the others.

Hawks and doves are the old symbols of the two sides in warfare. Today's debates rage over the use of excessive force (hawkish behavior) by police officers against minorities, or by roving bands of well-armed government thugs, or by drone strikes to kill small teams of terrorists in areas heavily populated by nonmilitary citizens. Doves focus on the victims rather than on perpetrators. Body counts stick deep in their craws, causing resentment that often morphs into a misguided wish for retaliation. But those who favor revenge, like the general public ignorantly searching for an enemy during an epidemic, easily forget that population wars are, and have been, eternal because they are fought over the timeless constraints of resource utilization: They conveniently forget that every living thing requires resources. And that eventually populations will come in contact with one another because there isn't a clean partitioning of each organism to a restricted patch of habitat. Resources commingle, and therefore populations are drawn together into a never-ending drama of coexistence.

Instead of an easy endgame to all these various population wars, it should be acknowledged that none of them are anything alike. Each interaction between populations is unique based simply on the extension of biology's most fundamental fact—no two individuals are identical. No two populations are identical either, and therefore solutions to population problems require specific plans of action. But, as will be pointed out, any plan will end in failure if the naive notion of vanquishing enemies is at the forefront of one's thinking. We will see that elimination of populations is difficult because of the one characteristic they all share: persistence. Populations are amalgams of previous populations. Usually a plethora of possible "enemies" from the past could be cited as participants in bringing about any given current population war. And therefore, simply not knowing who the enemy is could very well be our biggest failure in promoting warfare as an answer in the first place. Success will come from one tack only: long-term management of populations.

I began to appreciate population wars as a dominant theme in my thinking during my schooldays spent as a punk rocker. One of the common themes found throughout the punk subculture is a narrative of "them vs. us."[4] In Los Angeles, from around 1979 to 1981, it was very common for police violently to raid punk rock concerts, in full riot gear, batons swinging, because they believed that we were somehow destroying the fabric of youth culture and the hope of a clean-cut, organized future. The cops were spurred on in their disdain for us by songs that we sang in unison with our favorite bands: "California Über alles," "No Values," "Wrecking Crew," "Fuck Authority," and "I Don't Care About You."

I had no intention of overturning the status quo, simply reeducating it. I believed at an early stage that the "us" in punk referred to people who also felt like outcasts in the Southern California youth culture. Instead of nihilistic destruction, we believed in challenging the commonly held assumptions about what it meant to grow up in Los Angeles—that the only way to be cool was through surfing, skateboarding, hippy music, and smoking pot, and, as adults, the only way to be happy was to be rich so you could afford to buy more free time to surf, skate, and smoke pot.

I achieved early success as a songwriter, singer, and front man of Bad Religion. But somehow I never believed that it would last. So I remained committed to an intellectual pursuit of blending music with more "mainstream" academic work at universities. I immediately gravitated toward biology because of its focus on populations. All the fundamentals about natural selection and evolution—the foundations of all biology—require some sort of modeling about ideal populations and their interactions. This appealed to my identity as part of a subpopulation of outsiders—punk rockers in LA—and offered a promising way to make intellectual sense of that experience.

Biologists don't actually measure entire populations in the wild; they sample them, and extrapolate the data to make theoretical predictions about how populations evolve and what brought them to a particular point in time. I learned that it wasn't easy to idealize a population. I knew early on that punk rockers were more interesting than the stereotyped view that pigeonholed us as a population of kids who were useless no-goods, contributing nothing to our society. My punk friends in school were by and large the most intellectually motivated kids in the class, but the teachers by and large saw them as a subcultural aberration whose presence disrupted the tranquillity of the learning environment. These same individuals have gone on to achieve great success as adults—some of them are now both my friends and colleagues—with jobs such as academics, owners of media companies, movie directors, and record producers. I was drawn to using evolutionary themes as a songwriter, and mining the data of evolution to motivate my worldview. The story of any species' ancestral lineage is multifaceted and surprising in many ways, just as human groups—such as punk rockers—don't neatly conform to a predictable stereotype. Extrapolating from this realization leaves us with a profound conclusion.

Neither reprobate individuals, nor evil groups, nor species of predators, nor pathogenic microbes can be eradicated completely from our daily lives. Such a program of elimination will fail in the long term because of the persistence of populations. Rather than annihilation, it is tolerance and stewardship—providing resources and allowing some degree of freedom—that are required to coexist with other populations through time.

I find a lot of parallels in science and music, and I've dedicated my life to studying and practicing both. Music is what I do professionally. When I sing and play a song, there's no ambiguity, no frustrations, just clarity in those moments of joy. I wish I could say the same for life science. Despite the huge compendium of accumulated biological data in the past one hundred years, we biologists only flatter ourselves with the belief that we are any closer to a unification of the field than we were in the last century. The field is more diversified than ever before, but this needn't be a detriment. Rather than pretend that one unified theory underlies all of life science, I prefer to see the great accumulation of data as a broadening of the palette for creative scientific interpretations. Interdisciplinary syntheses often produce intradisciplinary squabbles, and that's usually a good thing for all the participating academic fields. New knowledge comes from such frictions.

Punk tradition also has a deep current of philosophical friction associated with it, and I tend to write new songs by using the same hardened punk themes but expressing them in new ways. By doing so, I believe new standards can be achieved. By the same token this book is not the proposal of a new principle or theory; it's more like a song, like an artist's attempt to craft a worldview. Using the facts of evolutionary biology and geology as my palette, I see parallels in the way populations interact with one another today and in the past history of life on earth. I see consistent repetitions in these interactions throughout the episodes of history, and I think they are trying to tell us something. I want to elaborate some of my findings, knowing all the while that, like a punk songster, some of what I have to say is distasteful to many.

Even though I'm not a household name like a lot of rock stars, I feel privileged that I regularly travel the world to perform. I've used this privilege to visit museums, head out on excursions away from some of the exotic cities we play, and come home, reinvigorated by my travels, to lecture at universities on topics in natural history such as geology, biology, and anthropology.

These great pillars of knowledge used to be more closely allied than they are today. The first geologists were essentially creators of Earth's chronology, using fossils and rock layers to mark the events leading up to the present day. Fossils had to be compared with skeletons and remains from living species. So every geologist had to know something about biology. Likewise every anthropologist had to understand biology—for humankind, the subject of the discipline, is a mammal of the Order Primates—as well as geology. The chronology of our species is ultimately calibrated by dating the layered sedimentary deposits associated with human activities. Human history, a story that is at least two million years old, is therefore superposed on a geological foundation.

Of course not all anthropologists today pay much attention to geology—for modern culture came about during just the most recent tiny fraction of time in earth history. In fact today scientists tend to specialize and become hyperfocused on one unique topic within their field. It's generally not encouraged at universities to mix disciplines, but rather to pick only one, and in it, find some species or patch of Earth and study it completely. Become an expert on something small—that way you cannot be assailed as a phony or a "soft science" advocate (a somewhat pejorative term usually reserved for those of us who try to meld disciplines at a coarser scale of analysis). The tunnel-vision approach to science has never appealed to me. I feel that too much focus might lead me down a long, lonely path of overspecialization, with no significance to a broader audience. Instead I'm motivated to arrive at a satisfying plateau of interdisciplinary conduct that blends, rather than excludes, other influences in my work. Topics from other fields, like influences from other genres in songwriting, have been the mainstays of my creative life.

Therefore, in science I took the "long way home." I got an undergraduate degree in anthropology, a master's in geology, and a Ph.D. in zoology, as I continued to write songs with my band and tour whenever the opportunity came along. It took me about fifteen years to accomplish this education; others have done it much faster.

Even though I still learn new things every day, I've started to be more confident in melding my classical training in zoology and paleontology with observations on the current state of our industrialized species. In a sense I've taken an anthropological approach, and I've come to recognize that the great disparities in quality of life today seem to be entrenched, not something that can easily be overcome simply by a person's hard work or dedication to a particular ideology. Rather it seems that the human population is participating in an elaborate ballet arranged by an unseen choreographer. Historical and economic circumstances have relentlessly thrust groups of people into contact with one another, bringing about violence and warfare. These violent episodes have repeatedly resulted in assimilations of the human species. Think of the conquest of Mexico or, as will be illustrated in a later chapter, the European settlement of New York. In both cases economic (or ecological) necessity drove populations together, violence ensued, assimilation resulted.

Such mainstays of the human drama are paralleled in populations of wild species. The easily observed historical pattern of conflict, followed by assimilation of populations brought into contact by the same environmental parameters, depicts what I call population wars. The first half of the book illustrates some of these. For instance, the rapidly expanding compendium of knowledge about other species that live inside us and all around us provides an eye-opening set of reasons to be concerned about our own population. As I lay out these examples, the perspective I'd like you to take is that of a naturalist, observing the empirical evidence of the wild, while still recognizing that we are a part of (not apart from) this elaborate biological pageant.

The second half of the book will require the reader to wear a different hat—that of the philosopher. The ultimate goal is to facilitate a shift in focus, away from the individual human being as an inheritor of free will from on high and toward an appreciation of the constraints placed on her by the circumstances of the group(s) to which she belongs. The result should be a recognition that we are all survivors of the population wars of the past, and we can use this knowledge to justify a peaceful stewardship of the planet.

Our journey will begin in the earliest days of life on our planet. Through tales of mass extinctions, developing immune systems, ancient human wars, the American industrial heartland, and our ever-degrading modern environment, the common thread of population wars will emerge. Our cast of characters ranges from simple cells in the primordial soup to plague-infected fleas in the Dark Ages, to American Indians in the Revolutionary War (1775–83) and modern-day, out-of-work skilled laborers in former manufacturing communities. We'll look at species that are in the process of becoming endangered, and others, like the horseshoe crab, that have remained relatively unchanged for hundreds of millions of years. All these groups have commonalities that weave the narrative fabric of this book: They are composed of distinct populations, they have all assimilated to varying degrees, and they all have attributes of coexistence in the modern world.

I feel it a great privilege to teach at Cornell University. Like most colleges, it's a hotbed of liberal beliefs and hippy ethics. Many years ago I started to see the bumper sticker around campus that spells out COEXIST—with each letter produced from an icon of the world's great religions. It's easy to dismiss this statement as too simplistic, yet the philosophy of coexistence has stuck with me. I think it is time to subject it to an intellectual discussion. We are all individuals and (as we will see) composed of numerous populations of other individuals as well, but we are also part of a population. The owners of that bumper sticker probably identify with one of the various religions depicted by the symbols, but their basic point in advertising the word is, Can't we as individuals put aside our differences and just get along without violence? It's a wonderful sentiment, and as an individual I agree with them. But as part of a population, we are constrained by various historical contingencies. The humanitarian goal of the twenty-first century as I see it is to learn those constraints, face them squarely, eliminate them, and agree on a global course of action for our species. The goal of this book, however, is simply to make you aware of some of those constraints, understand how they operate, acknowledge their importance, and consider the validity of this approach. What you choose to do next is up to you.

Obviously, I love natural history and specifically the study of evolution. But I also have children, and I take my role as a parent very seriously. Part of that role is figuring out the best way to introduce them to our culture. This is a process of assimilation, and each child is like a population. I have to acknowledge that my kids will experience pain and trauma before they attain stability and self-sufficiency. Although as parents we try our best to ameliorate suffering, growing up is always painful. In a sense the painful changes kids experience throughout life can be slightly mollified by parental stewardship—coaching, training, leading by example. Most important to this effort, however, is the realization that a future awaits these innocent loved ones. By presiding over the development of their bodies and minds we prepare them for that future. We feed them food and ideas, in each case creating an environment to meet the challenges they will face. Whether we admit it or not, we are acting as ecological stewards for the most precious ecosystem of all—the minds and bodies of our offspring, composed of developing cells and populations of symbiotic microbes.

By extension of my instinctive drive as a parent to foster intelligence and health in my children, I propose that there is a fundamental reason to advocate for environmental stewardship for the planet at large. The kids will inherit the Earth we leave to them. Now that every inch of the globe has been explored (and I've had the great privilege of exploring so much of it), I'm convinced that humans have the technology and knowledge to leave it in better shape than we found it at the time of our birth. There are many steps to this ambition. Mapping, for instance, is a fundamental one that has already gotten off to a fantastic start—just spend a half hour on Google Earth. Searching for the basis of coexistence, however—the theme of this book—is another key factor in the successful stewardship of the future. After all, human coexistence ultimately depends on our coexistence with other species.

1

PERSISTENCE IN THE FACE OF EXTINCTION

The single most valuable thing any of us can do intellectually is to understand how populations interact and affect one another. Populations have a tendency to persist. It's futile to believe otherwise. The populations you seek to exterminate will most likely continue, whether it's dandelions on your lawn or enemy armies vying for land you covet or religious fanatics whose ideology you despise. Since none of us had any control over the circumstances that brought us together, the only viable way forward is compromise.

I'm asking you to take a coarse-grained approach to your interpretation of compromise in this book. We'll examine adaptation, and I will ask you to not see it as a "struggle for existence" but rather to acknowledge that populations become altered when they coexist, and eventually reach a state of relatively benign equilibrium. With humans, I will insist that rational communication—diplomatic compromise—has to take center stage in order to avoid violence and bloodshed. Through it all I hope you will be left with an appreciation for how persistent most population phenomena are, and how important history is in explaining coexistence today.

As a starting point we will consider our own nation's history. It's depicted in a pretty straightforward manner in elementary textbooks, usually beginning with the Battle of Lexington, or the Boston Tea Party, or often the first Thanksgiving, where tales of overcoming adversity led to strong moral character and good citizenship. Nearly every American schoolchild has heard of such heroic figures as Captain John Smith, George Washington, and Thomas Jefferson. These are convenient tales, because they "leave out anything that might reflect badly on our national character." Since childhood we've been reminded: "You have a proud heritage, be all you can be, after all, look at what the United States has accomplished."[5]

Overly simplistic introductions to our nation's past might play a role in grade-school education, but America's history can be understood only as a complex saga of populations from disparate backgrounds. What's easily sold to youngsters as a straight-line evolutionary development of our citizenship, from wilderness settlers to urban captains of industry, is in actuality much more highly branched and complicated when we consider the various populations that came into contact historically.

I'm not old yet, but I'm getting there, and one thing has become abundantly clear: I am more a result of previous circumstances than I am a fulfillment of youthful dreams and willpower. The idea that we are the end product of a historical unfolding of events rather than self-made entities who exercise total self-control over our own destiny is offensive to some people. Successful people tend to believe that they acquired their status and fortune through hard work and wise administration, period! Sometimes I hear "rich and famous" people hailing themselves as wizards, geniuses, or miracle workers without ever acknowledging or considering other people's roles in the equation. This attitude can be characterized as self-important. The more realistic among us are keenly aware of the people and past circumstances that aided in our own success.

Have you ever considered that your station in life—not only your physical whereabouts but also your social, emotional, and economic well-being—is beyond your control? If you're like most people, this possibility sounds offensive, incomprehensible, and alien, simply because you are so confident that everything you have in life came from hard work and intelligence, and is therefore well-deserved. This book will perhaps help you see otherwise.

When I consider my own journey, I think back to how I ended up living here, in a rural part of upstate New York. Sure, I came here for college, but I never thought I would stay. After all, my grade school was in Wisconsin and my high school in Los Angeles, both places I love that filled me with a sense of modern American identity. I went to college at UCLA, only a few miles from Malibu beach. Who would leave the social and climatological paradise of Los Angeles for the stark harshness of upstate New York? Since my professional career is centered in the entertainment industry, I've never fully left Los Angeles. It's the headquarters for my musical identity, and I still spend a great deal of time there each year. But the latest chapters of my domestic life have, for a quarter of a century now, taken place in a region within a region—the southern tier of New York State in Finger Lakes country.

Most people would consider this area distinctly American. It is the land of cheddar cheese, apples, beer, hardwood lumber, forests, salmon fishing, deer hunting, baseball, NASCAR, organic farms, and wineries by the score. But three hundred years ago upstate New York was British territory. The English monarchs, governing from London, made all decisions about village politics, farming practices, and relations with native peoples in the New York territories during the seventeenth and much of the eighteenth century. Upstate New York villages were laid out as old English hamlets were, a cluster of houses and shops surrounded by agricultural fields. Precolonial foundations still survive in the center of some villages. These stone foundations would have supported the millhouses that were powered by the waterways that served as the lifelines of the communities.

Before the English claimed it as their own, France made serious attempts to settle upstate New York as part of their effort to create "New France." The French built numerous forts there, including Fort Niagara in 1678, and Fort Saint Frédéric in 1729. The decisions made by the kings of France during this time were geared toward establishing a permanent presence in the New World to facilitate three main goals: garnering the fur trade, planting Christian missions, and blocking English expansion in North America.[6] Many Europeans came to America as workers on ships or builders of forts, and ended up settling near these remote American redoubts.

The French and English are usually introduced to American schoolchildren as the vanquished (French) and the victorious forefathers (English), but if we go back even farther, we will see that the French had predecessors who claimed this area as their own. This land was settled by Native Americans of the Iroquois Confederacy. In fact, when Europeans first visited upstate New York, the Iroquois league of Five Nations (Seneca, Cayuga, Mohawk, Onondaga, and Oneida, and later a sixth, the Tuscarora) had well-established towns with buildings made of lumber, agricultural operations, trading enterprises, and fortifications. Though often on the move, and relatively sparse in population, the Iroquois people were the first people in New York to build houses, establish trails and roadways between towns, and clear fields for agriculture. They controlled the headwaters of all the most important water trade routes in the Northeast (the St. Lawrence, Delaware, Allegheny, Susquehanna, and Hudson Rivers). In short, they were a well-organized nation before European explorers made contact with them.

The historical trajectory from what has been called Iroquoia to modern America was not a sequence of extinction and replacement. None of the preexisting populations of upstate New York were vanquished. Their descendants are still here today, their numbers continue to increase, and their story gets ever-more intriguing with each passing year. The cultural landscape today is in fact a complex mélange of past populations. Archaeologists continue to excavate Iroquois towns throughout the state.[7] Many of today's major thoroughfares and state highways are paved over the ancient trails and trade routes of the Iroquois. Many modern agricultural fields in the region are still used for corn, as they were during the historical period of Iroquois life.

This land of which I am now a resident is, therefore, a mosaic of past events that have come together to form inescapably the historical fabric of every community in the upstate region of New York. My adopted home contains the indelible stamp of previous populations that no longer predominate but yet still persist. In many ways this ad hoc unfolding of history is similar to my own story of wandering and settling down.

I ended up here to pursue graduate studies, at least in part because of a family tradition that favored land-grant universities. Cornell is the New York version of those types of schools; publicly funded, a wide breadth of general studies to choose from, and a liberal spirit of education for every citizen, no matter the race, economic background, or creed. These were the values that my family taught. The congressional acts that allowed those opportunities to be put in motion, however, went far beyond my own family to circumstances that were set in motion long before I was born.

In 1860 federal land grants were doled out to build universities all over the country; Ithaca was chosen as the site for the New York State College of Agriculture and Life Sciences, which, with a grant from businessman Ezra Cornell,[8] became Cornell University. Ithaca has been the premier college town in New York ever since that time. I moved here in 1990, and I've seen a lot of changes in the region. It now has all the trappings of modern consumerism—big-box retailers like Home Depot and

Lowe's, Target and Walmart, retail malls, movie theaters, and excellent grocery stores, yet the entire town has only roughly thirty thousand people. When the schools are in session (Cornell University and Ithaca College), however, the city swells to double its population.

Ithaca is surrounded by smaller hamlets that are connected to one another and the bigger cities by two-lane highways. Between the retail and college meccas such as Ithaca and Elmira, and larger cities like Rochester, Syracuse, or Buffalo, are scores of villages and towns with very few services or industries. In fact most people who visit here wonder what could be the reason for so many tiny settlements. Living here just doesn't seem to make sense to the average person. Doesn't there have to be some reason for people to exist in these tiny, out-of-the-way Podunk towns?[9] But this brings us to one of the key points of this book: Most of life's phenomena depend on historical causes—people end up where they are, such as in small towns or out-of-the-way hamlets, because of historical factors that affected previous generations.

Modern culture tends to judge "country people" for not taking advantage of the conveniences of city life. The only rural families that regularly make it on TV (usually reality TV) are portrayed as being almost laughably simpleminded and backward (though often with some kind of redeeming "heart of gold"). However, I've come to the conclusion that it's foolish to view rural residents in comparison with their urban counterparts in modern America. Rather, I see them as part of an unfolding of regional history that stretches back to the years in which these lands were settled.

In fact I am a convert; I came to appreciate rural life after thirty years of living in various cities. I have been able to recapture a tradition that my grandparents enjoyed: living in small towns surrounded by rural landscapes. Today, thanks to modern technology, building materials, transportation, and the New York State electrical grid, my family and I can live deep in the heart of rural America with all the conveniences that modern life affords. Our house may be in the middle of an alfalfa field surrounded by thirty-five acres of hardwood forest, but we still have a fiberoptic Internet connection. Many of my friends back in the big city are jealous. They can't get fiberoptic in their own city dwellings because it's a newer technology that requires a lot of infrastructural reconfiguration. In the country, however, there is plenty of space to put in new technology. In the crowded city it's not so simple. Most of the space is already encumbered by the cable wires of older technology. Because of this, fiberoptic hasn't yet pervaded every nook and cranny of the major cities—a good demonstration of historical constraints on "progress."

My fascination with history also makes it exciting to live here. I step off the porch and walk to the forest edge just a few yards south of our house. Entering a trail, I observe a myriad of hardwoods, maple, oak, black walnut, cherry, and ash, among others. History is never far from my mind as I witness an occasional sassafras tree or yellow poplar, both southerly species making their way northward since the retreat of the glaciers left this region and it grew progressively warmer in the last ten thousand years.

Farther on, down the trail, our property line ends at a rushing creek. Its erosive action began as the last glaciers melted away and retreated northward. The creek flows over bedrock that is ripple marked and laden with fossil marine invertebrates, signatures of a near-shore marine environment in a bygone era in Earth history, 380 million years ago (mya). With each passing storm the creek erodes more of the bank and reveals a slightly different topography, and new areas of ancient bedrock are exposed for examination of new fossils. How these ancient creatures—revealed as fossils—coexisted drives my curiosity, and drives me to search for parallels in the way the trees and forest creatures coexist today.

But there's more history to discover in this region as well. The receding glaciers left us something else besides the forest species and carved-out bedrock. Slightly east of our house, and down a significant declivity, is a bog, full of black mud and sticky as molasses. This is the result of slowly percolating, oozing water, trickling from the hillside, bringing with it only the smallest particles of mud and clay. It is in these Pleistocene[10] bogs that giants of the past have been trapped and preserved as fossils. The mastodon, *Mammut sp.*, an elephant that stood nearly eight feet at the shoulders, made this land its home. In fact, only fifteen miles from our property a farmer discovered a complete skeleton on his property in a similar bog deposit. This region has become famous as a type-locality for the mastodon, which means that it contains numerous bog deposits that are favorable for the preservation of this grand elephant species of the past. I don't have the heart to bring in huge excavating machinery and dredge the bog on our property. I fear too much damage would occur to the forest and beautiful scenery outside our windows. But I always hope to find, on one of my almost daily hikes, some elephant tusk or leg bone eroding out at the edge of our bog. Because I am sure that whatever died down there in the last ten thousand years is still buried in the mud. The allure of adventure, driven by the surrounding natural history, leads me inexorably to a meditative contemplation about how this natural setting came to be, and the path I took to find it.

The woods out my back door are a long way from my day job as a singer/songwriter in Bad Religion. But they serve as an escape and a recharge every time I come home from tour. Our band is still headquartered in Los Angeles, and thanks to modern airline travel I go back and forth between New York and California at the drop of a hat. Flight crews know me as a regular commuter. I'm on the road at least three months a year with the band. I'm in the studio, recording in LA, for a couple of months every two years. And I'm back in Ithaca teaching evolution at Cornell in the fall. It's a busy but highly rewarding schedule. I'm very fortunate and pleased at the way my life has turned out. But I can't honestly admit that any of this was predicted, or even predictable.

I was born in Madison, Wisconsin, and within a couple of years settled in to our family house, still occupied by my dad today, in a quiet neighborhood of Racine, Wisconsin. Like most suburban areas, every square inch of the landscape around my childhood house had been shaped by human hands. The street where I grew up playing touch football and baseball is made of poured concrete underlain by a foot of crushed and sifted gravels. The curbs are smoothly shaped and molded by workers with tools that are now long obsolete. Each property is edged by a narrow strip of grass bordering a perfectly straight stretch of sidewalk dotted by ornamental trees, one for each house along the street.

It can be hard to imagine that suburban Wisconsin was once virgin prairie. When the developers arrived in the 1940s they dug tens of thousands of pits into the native prairie soil. Each one eventually became a basement romper room like the one in which my brother and I spent countless hours playing Ping-Pong and pachinko, listening to pop music, and shooting darts. The basements were lined with cinder blocks and finished with a skim coat of cement mortar to form the foundations of tens of thousands of neat little houses, each with two bedrooms, a living room, a bathroom, a small kitchen, and a one-car garage. Our house, like all those surrounding it, is a testament to the postwar era when young couples could affordably start a family and fill their home with American-made consumer products and agricultural goods. These landscapes might seem as American as apple pie, but they don't resemble anything that the American Indians would have recognized. The prairie today is plowed under or paved over. The lawns are landscaped with nonnative grasses, and the flower gardens filled with annual flowers from Europe, Africa, China, and the Western United States—all of which are aliens in this transformed ecosystem.

Hardly anything of the original habitat remains in this small suburb. Even the parks where we used to play ball, the open spaces at the nearby school playground, and the beaches of Lake Michigan only a quarter mile away are cultivated and maintained by unnatural processes. The Parks and Recreations Department has to work constantly to maintain the illusion of a natural world. They weed out the native species and plant cultivars in their stead. They fertilize and water the lawns to keep them lush through the driest summers, and spray herbicides to keep the dandelions and crabgrasses in check. Even the beach is artificial; the city brings in sand to make littoral areas more pleasant for picnics and the Fourth of July. As kids we could spend all day outdoors, riding, playing ball, going to the ice-cream stand, and so on, and never come into contact with a single piece of native prairie vegetation. Our bike rides to the beach, and over the endless miles of bike "trails" (on top of railroad dikes and ballast) traversed a completely manufactured ecosystem. It was, and remains, as most populated areas do, a finished, human-made vision; the product of a set of blueprints, laid out by postwar city planners and dreamers of the American future.

By the time I was eleven, I had moved with my mom to Los Angeles, California. Physiographically, the San Fernando Valley shares very little with Racine, Wisconsin. Dominated by high surrounding mountains, LA is a patchwork of flat basins crammed with humanity. But the American dream of neat little houses in fully finished landscapes translates as perfectly to Southern California as it does back east. In every corner of our country, regardless of biome, tectonic setting, or natural history, the concept of creating a controllable refuge from wild nature persists.

In LA our house sat on a street with curbs, narrow plantings along the sidewalks, and ornamental trees and flowers from all parts of the world, just like back in Wisconsin. We had an avocado tree and an orange tree in the backyard, which seemed both natural to our new home and foreign to our Midwestern experience. But these, too, were mere figments of a developer's imagination, for they are of Asian/South Pacific (citrus) or Central/South American (avocado) origin and were cultivated by immigrant growers who themselves were foreigners in Southern California.

Within the twenty-thousand-mile labyrinth of roads that circumscribe the neighborhoods of the San Fernando Valley and Los Angeles Basin is a seemingly unending patchwork of cultivated plots, most of them less than a quarter acre, made of poured concrete, trucked soil, sifted gravel, and nonnative plants. Most of our California neighbors had garden and lawn landscapes that resembled those of our neighbors in the Midwestern United States, using slightly different plants, and needing far more irrigation. It was, like all the neighborhoods across the country, a finished plan, with no consideration given to the preexisting conditions and how they might affect the sustainability of the community.

American suburbia is no different from the metropolitan sprawl of other countries around the world. Modern humans have come to idealize cultured society as one that has control of the wild. Paved streets, curbs, and sidewalks, like the finished woodwork of the house interiors, are symbols of the good life. Wild species, rough, rocky outcrops, and native vegetation are obliterated or moved aside to make way for the vision of both developers and homeowners.

I have a love-hate relationship with the suburbs. I grew up in the suburbs, and they still feel like home to me. I like the convenience of sidewalks and the closeness of neighbors, some of whom lived right next door and became lifelong friends. I couldn't imagine living in a high-rise urban apartment without a lawn to play ball on or a garden plot to grow vegetables in. But as I grew more worldly, and began to explore wilderness areas more, I grew disillusioned with the concepts of city living, both suburban and urban. I realized that the dreams of postwar city planners were not sustainable on a large scale. What might work in Racine (1970 population around 95,000) would never work in urban populations that were mushrooming to the size of those now in Los Angeles, New York, São Paulo, or Tokyo.

Something else was happening too. My college years were spent doing fieldwork in remote places as training for my geology degree. Backpacking and camping trips, from weekends to monthlong expeditions, to deserts and rainforests began to seduce me. I started to doubt the wisdom or practical realities of living in the city, and to feel a deep longing to live a life that was closer to the wilderness. Out there in the field I wasn't constrained or controlled by traffic, concrete, and millions of other people. I realized that I was at my happiest, and felt the most emotional satisfaction, living in unfinished landscapes. A forest edge, a river's rocky lag deposit, a crumbling outcrop, a muddy bog, a wild meadow—all these things came to dominate my ideal of beauty, and I believed that they could be appreciated most readily in a rural domestic setting.

I'd already spent a large amount of time in wild landscapes that were untouched by humans. My studies had taken me to the Amazon and to the most remote wilderness areas of the Lower forty-eight. During graduate school I worked in the mountains of the Sangre de Cristo range in Colorado, the Bighorn Mountains of Wyoming, and the lonely Chiricahua Mountains of southeastern Arizona. I had visited boreal forests of Alaska, desert valleys of California, and the dry plains of the Altiplano, high in the Andes. These places, in complete contrast to the American suburbs, were "unfinished." They were raw and—in my eyes—filled with promise and possibility. In my youth everything I laid my eyes on was finished by human hands. Someone sitting in a Parks Department office had decided on a plan of action, from the acceptable height of the grass to the color of the flowers in the municipal park. In wilderness everything I saw was touched only by the ongoing processes of weather, and unending forces of geology and biology. "Wild" became a synonym for "unfinished." And I came to believe that unfinished was a much more exciting and authentic way to experience life than the false sense of tranquillity that came from the suburbs.

Eventually I settled in a place that has the perfect blend of wild and modern. The towns and villages of upstate New York inspire a constant regard for populations and their histories. This region seems the ideal mixture of nature, remoteness, and fancy amenities. There is so much that is unfinished here. The forest is still populated by species that have been here since the Ice Age. The geology of the region is visibly exposed in outcrops of 380-million-year-old layers of shale in every gorge, hiking trail, and road cut. And the human population hasn't expanded here, so the neatly cultivated suburbs, like those seen surrounding America's largest cities, have not replaced the colonial village. And yet the cities nearby and Ithaca itself provide the best things in modern culture: great concert venues, fine cuisine, foreign films, museums at world-class universities, and public radio stations.

There are daily reminders of the difference between urban and rural life here. On the one hand, "urban" suggests finished plans, industrialized efficiency, and clean edges. In the city we can hold on to the belief that we have dominated nature to conform to our needs for convenience and high culture. The past is skim coated to give the illusion of renovation. Nothing is left of the original landscape in the "concrete jungle," and one is hard pressed to find rocks, animals, or plants in their native state.

"Rural," on the other hand, implies a mosaic of historical remnants that to many are easily forgotten, but are in fact merely lying in wait for the modern age to notice. Whether it's natural resources for industrial exploits or charming country folk for the airwaves, time and again modern citizens turn to the rural for their needs. The reason for this is that, perhaps secretly, we all see the prospects for the future in the raw materials of history.

One of the strongest emotional outcomes of rural life is belief in ecological stewardship and a longing to care for the future. This is because so many rural citizens live much "closer to the land" than do people in urban places. Farmers, land managers, and natural-resource administrators are in much higher concentrations in rural townships and villages than in big cities. These people, in particular, are keenly in touch with the latest science and technology, particularly as it relates to ecology. Far from being outdated Luddites, the people in most of the rural communities I've visited embrace new discoveries in wireless technology for mapping and communication, high-efficiency vehicles, and genome research, for example. Most of them use the latest manufacturing innovations in each of these fields. Furthermore they depend on a culture of sustainability, not just for the next harvesting season but for maintaining their operations and ways of life indefinitely. Whether it's a farmer overseeing his fields and annually rotating his crops, or a timber manager selectively harvesting a managed woodlot, life in the "country" implies a need for constant stewardship, a harnessing of weather, geology, and biology. These stewards understand that long-term management is the more fruitful mind-set rather than shortsighted pest control.

When we attempt to tame nature for our own benefit we ignore the fact that all the millions of populations that make up the natural world are constantly in flux. Nothing is ever "finished." The only thing that can stop the process of evolution—or at least the possibility of evolution—is extinction. Yet there is something about our suburban and urban world that seems to exist in opposition to this reality. We humans like to imagine that we are the "end of the line," the ultimate omega species that will never be improved upon. But the reality is that we are just like all the other species, a result of history. There is no ultimate plan or goal. When

I look at the suburban world I see evidence of our very human unwillingness to accept that fact. As a kid I didn't question what those thousands of Wisconsin dads, mowing their lawns, or the local park authorities dictating which ornamental shrubs edged our playing fields, were really trying to do. Now I realize that, in the grand scheme, they were attempting somehow to justify their existence by constantly keeping the wilderness at bay. They were attempting to demonstrate that their creation—a modern human paradise—was "finished." But it was all an illusion. They were merely creating a new theater for population wars.

I can't blame them for their commitment to their purposeful cause. After all, the consideration of populations makes solutions more difficult, whether you're writing a history of the United States or attempting to maintain a property. When one has only the weekend to do chores before the next workweek begins, the simplest solution takes precedence, rather than the healthiest long-term option. For instance, most people who have houses also have plantings and lawns to tend. In general they want to solve the problem of weeds and pests that inhabit those lawns, shrubs, and trees around their property. Their natural choice is pesticides, herbicides, and rodent traps. This usually results, however, in only a temporary decrease in the pest populations. Soon the survivors—or the immigrants from a less-diligent neighbor's yard—reproduce and the population grows to nuisance level again. The sale of these antipest products depends on the deep-seated belief that populations can be exterminated, vanquished, and the problem will go away. The problem of pests and weeds comes right back next year because the longer-term solution—managing the populations through time—is much more difficult.

I have more opportunity to observe populations of weeds and pests than does the average suburban homeowner. The plot where we built our house is a clearing in an old agricultural field adjacent to acres of woods. Nearly every day I spend time watching native animals—mammals, birds, reptiles, insects, amphibians—as they go about their routines, much the same as they have for centuries. I am aware of the invasive species as well—the unmistakable signs of human interference—starlings brought over from Europe, and honeysuckle shrubs along the forest edge. I try to maintain the forest by removing invasive species. I have no problem shooting starlings, and every spring and summer I spend many days attempting to obliterate the honeysuckle from my land. It's an almost hopeless endeavor—the honeysuckle is unbelievably tenacious. I use my tractor to pull out the bigger bushes, but the smaller saplings need to be pulled out by hand. Here and there I see species interacting, robins scarfing larvae from the spring soil, mating pileated woodpeckers chasing each other around the trunks of tall hardwoods, and I am satisfied that most species are coexisting as they have for millennia. But the invasive species' populations have to be culled in order to maintain the livelihood of the native species. I know that I cannot exterminate them completely, but perhaps if I dissuade them from finding safe havens on this property, and other citizens do the same, a healthier ecosystem will result.

Our property has multiple, distinct microenvironments: We have hardwood forest, bogs, and a creek. Primitive grasslands edge our creek, and an open meadow surrounds our house. I particularly like the small stands of beech and black walnut trees interspersed within the larger forest. It all looks incredibly healthy, but here is the secret about forests: A dying forest can look as lush and as verdant as a healthy one. Our hardwood trees are constantly under attack from vines, beetles, and overcrowding. In addition to the honeysuckle culling, I spend a lot of time chopping down Asian thorn apples, pulling down grapevines, and observing the telltale signs of a fresh beetle infestation in a newly mature beech. A healthy forest provides shade that retards the spread of sun-loving invasives; this is why the edges of disturbed forests (for instance where the forest has been cut down along the edge of a highway) are often drowning in thick growths of invasive vines while the shaded heart of the forest is clear and unencumbered. The intense sunlight available along roads and in disturbed clearings is the necessary ingredient for these sun-loving invasive species.

I think of myself as a steward of the environment; my property is healthy, but it takes a huge amount of effort to keep it that way. If I stopped trying to control the invasives, this plot of land would soon start to degrade despite my stewardship. In a few years the fragile, native ecosystem would be overrun by aggressive outsiders. Yet I am still aware that in the long run this forest will change; the populations will continue to mingle. Even if I could destroy all the honeysuckle on my property, it still wouldn't change the fact that honeysuckle is incredibly successful and a (most likely) permanent addition to the Eastern landscape. So I work within the reality of my environment, seeking to maintain a functional compromise between the various species under my domain. In short, the acknowledgment of all these interacting populations results in an awareness of the need for stewardship and a humble admission that my efforts require help from other like-minded people.

The natural forces that shape our world are more powerful than any human individual. The creek out back on our property winds its way across an ever-changing landscape of gravel, sand, and soil. A heavy rainstorm will completely reshape it. The cut banks erode, taking with them the sediment and the overhanging trees, while the point-bar deposits broaden to form pleasant forest "beaches" where we can sit and enjoy the churning of the rapids in the channel. Next season we might have to move our vantage point because the stream channel is ever changing, always modifying the habitats through which it cuts. Its power to carve new paths, and create new overbank areas, is totally dependent on rainfall and snowmelt; a very snowy winter will have changed the river dramatically come springtime. The work of nature is never done here; it dominates the rural landscape, and is a constant reminder of past worlds upon which this current one was built. Our land is literally "unfinished," and always will be. There is no endpoint where it will be considered complete. The individuals and populations that make their lives on its ever-changing geology will be forgotten soon enough—myself included. We are all bit players in a larger drama that has no story line and no conclusion—it may sound nihilistic, but I find this oddly comforting. I spend many meditative hours thinking about the populations who lived out their brief lives on what is now "my" land. And this leads inevitably to extrapolation about all human beings. We are a small part of an ongoing interaction between countless populations from multiple species, all of which are simply seeking to live and reproduce. I am absolutely certain that there is no ultimate purpose in life, only the proximate purpose of seeking to live it as well and as meaningfully as possible. Part of this involves the ethic of respectful stewardship of other species. At the same time, however, I am compelled to view our own species as the most important. More on this later.

The rural world is constantly in the process of becoming something else. In many ways urban life has this potential, too. But in my experience urban life is also full of failed experiments, and this depresses me. Abandoned apartment buildings, unused warehouses, broken glass on the facades of dormant and crumbling factories, and rusted-out vehicles in long-abandoned loading docks—these are the things that stand out to me when I visit the urban areas of this country. I feel left out, as if a plan was made for a previous time and I came along too late to enjoy its heyday. All that's left, it seems, is a population trying to get by amid the forgotten buildings and industries that used to be so vibrant and important to its sustenance. The citizens are apparently trying to forget the past while attempting to figure out a new way to make a living amid the crumbled infrastructure. But the defunct factories, rusted railcars, and unhealthy conditions keep fueling an ever-present hopelessness.

There was a time when the urban environment thrived in America, when manufacturing industries were the mainstays of their communities. Cities like Buffalo, Cleveland, Detroit, Toledo, Chicago, and Milwaukee were full of people who worked at factories, and their families were busily living out their lives in decent neighborhoods adjacent to the industrial areas. Beginning in the late nineteenth century, iron mills, steel refineries, fabrication plants, die-cast operations, electronics and electric household appliance manufacturers, printers, agricultural equipment makers, car parts factories, and many other companies employed hundreds of thousands of workers in these cities. The factories and manufacturing and assembly companies were so numerous that tens of square miles in each urban area were devoted to their plants, and they were connected by rail to other industrial centers throughout the United States. Workers, and their many hundreds of thousands of family members, depended on these companies for their nearly constant supply of job openings and high wages. The families lived nearby in planned communities that were to become suburbs, while fathers earned healthy livings that allowed them to commute home each night and to pay the bills for the family's modest lifestyle.

I was born into this world in 1964; it was a pivotal, turbulent year, one that marked the transition between the relatively peaceful and conservative postwar era and the beginning of a more modern, volatile America. In the same year the first Ford Mustang rolled off the assembly line in Detroit and the genetic code was all but deciphered in Marshall Nirenberg's lab in Bethesda, Maryland. Bob Moog, a graduate student at Cornell, presented his new invention at a small gathering. It was an electronic music synthesizer, the first of its kind, that proved music could be produced purely by electrical circuitry. And, tragically, three youngsters (one a Cornell alumnus) were murdered by racists in Mississippi for their work in registering black voters in that state. The national outrage caused by this racially charged incident led to the passage of the Civil Rights Act, also in the year of my birth.

Much of the country's working population had jobs in manufacturing in 1964. The heartland of industrial manufacturing at that time was in cities on or near the shores of the Great Lakes. Chicago, Milwaukee, Gary, Detroit, Cleveland, and Buffalo all had easy access to shipping. Railroads could transport machinery to the ports of these cities, where the goods were loaded onto cargo ships that entered the Great Lakes and made their way to the Atlantic Ocean via the St. Lawrence Seaway. The world market was only a shipload away. Alternatively, these cities had easy access to the Ohio and Mississippi Rivers, the gateways to most of the American population.

Factories, foundries, grain mills, and warehouses made up the largest portions of the job sectors throughout these cities. In Wisconsin, where I spent my youth, my parents were academics, but most of my friends had dads that worked at factories of one kind or another. Our region was a center for agricultural machinery manufacturing; J. I. Case and Allis-Chalmers were the two giants—tractors, threshers, combines, plows and other implements.[11] But they also manufactured automobiles, hydroelectric turbines, cement kilns, municipal waterworks, and pumps. The list of American-made machinery goes on and on, and many of them are still in use in cities throughout this land. In fact, if you've ever been in Chicago, San Francisco, or New York, you can bet that Allis-Chalmers built a good number of the pumping stations that move the cities' millions of gallons of water, or the electrical turbines that powered the electric grid in the last century.[12] That these and so many other manufacturing products are still in use today is testimony to the good design and great endurance of American-made machines from these industrial centers.

My grandparents and parents thought their world was "complete." They had small but nice houses, good jobs, and social stability. Why would anything change? But in the mid-1980s something shifted that was long in the making: The domestic machining, assembling, and smelting industries took a nosedive. Financiers saw wasted dollars in the high cost of human labor. In favor of profits over the lives of the workers, most of the "rust belt" industries were deemed inefficient because higher profits could be achieved with cheaper labor in other countries.

Almost overnight the factories closed their doors, and hundreds of thousands of American workers were suddenly jobless. Most of these factory buildings, such as the Allis-Chalmers plant in Milwaukee, or the Delco Remy plant in Indiana where my grandfather used to build car parts, still stand today. Even though they closed their doors and boarded up the windows years ago, their buildings dominate the landscape because these facilities took up hundreds of acres of land. U. S. Steel, the main source of all that American metal, with headquarters in New York and its largest mill at Gary, Indiana (on the Great Lakes), was the number one producer of steel in the world. A strike in 1987, combined with a failed hostile takeover, and the subsequent shuttering of three factories, dramatically compromised the company. This caused a ripple effect in the U.S. manufacturing sector that resulted in tens of thousands of layoffs, and an evaporation of manufacturing jobs domestically. In that same year Allis-Chalmers announced that it was merging with a European company and that many of its manufacturing plants, including those in Milwaukee, would be closed permanently. Meanwhile, in my hometown of Racine, J. I. Case had merged with International Harvester of Illinois. These companies, and many like them, were complex and had numerous components. When two companies merged, the component sectors were often sold off in an effort to make the newly merged corporation more efficient. In this nationwide restructuring process, tens of thousands of working citizens were affected, and the corporate push for efficiency brought massive layoffs. Thousands of highly skilled machinists, builders, and their administrative cohorts lost their jobs overnight due to decisions made by the corporate boards of directors. In short, you could say that the 1980s and 1990s were an era in which America stopped making machinery and producing raw materials and started importing more foreign-made goods and materials. Today China makes most of the steel we use in this country.

The economic upheaval of the 1980s and 1990s left its mark on the communities where I grew up. Much higher unemployment meant more poverty and deteriorating neighborhoods. But even that wasn't as depressing as the stark devastation that beset the factories and industrial quarters of the cities. When you visit Racine, Wisconsin, today it is very similar to visiting Detroit or Cleveland or Buffalo. There are still neighborhoods near the old factories, but the people who remain are jobless and feel hopeless that they will ever be able to have a livelihood like that experienced forty years ago by their parents or grandparents during the now-long-gone age of twentieth-century manufacturing. The factories are boarded up and rusting, along potholed streets that stretch for miles, transected by abandoned railroads. Most of the stark edifices have weathered signs outside advertising "available space for rent." The houses in these neighborhoods have lawns and gardens that show signs of deferred maintenance. Some have plywood covering the window openings, or sagging roofs, or broken porch beams. Many of the people haven't left. After they or their parents lost those manufacturing jobs, there was nowhere to go. So they persist as a population, neither vanquished nor victorious, getting by as best they can in a landscape whose fate was sealed by corporate mergers thirty years ago.

In a lot of ways these modern Americans who have lost their jobs are similar to the American Indians who came before them, also disillusioned and cheated by policy decisions made in distant places. They are groups of people who helped to form the fabric of our nation. Although the world around them has changed dramatically, both groups still exist. Neither extinct nor thriving, these people live under the same laws and consume the same foods, watch the same mass media, and attend the same schools, but are subjects of a flawed narrative that I find pernicious. The narrative states that their station in life is determined by the choices they make, by their willingness (or lack of it) to compete, or by their mental and emotional fitness for modern life. These bogus stories stem from an incorrect reading of natural history and evolutionary biology.

A more realistic explanation is that anyone's station in life is due instead to preexisting circumstances that affected the populations to which they belong, conditions that are so powerful that only a significant amount of willpower or good luck plays any role at all in advancement. Therefore, rather than demand that these individuals mirror the life choices of people living in more affluent and successful communities, we should instead try to understand those preexisting circumstances, and work toward incremental improvement in those communities. As you will read later, I'm not a big believer in the significance of competition or free will; instead I think our chances of succeeding or failing in life are more constrained by extraneous factors, often generations before we are born. The truth is, a select few people made some very selfish, shortsighted policy decisions decades ago that drastically altered the future course of these populations. I think it is a lazy and often cruel rhetorical crutch to assume that individuals all have the same opportunity to "be whatever you want to be."

Amid all the industrial decay there are signs of populations persisting and even thriving. The Iroquois are slowly writing the next chapter of their people's history. American Indian groups on tribal land are financing and building impressive high-rise casinos and hotels. Some of these are set directly in the industrial parts of the very same Northeast cities in which their ancestors traded with white men during historic times, before the cities received their modern names. These enterprises have become tremendously profitable due to the popularity of gambling in America. Billions of dollars of revenue are generated in New York State alone. Casinos and "racinos" (horse-racing facilities that also offer gambling) are thriving from Milwaukee to Detroit to Buffalo to Syracuse, and numerous locations in between, like the southern tier of New York State. These updated gambling dens offer their customers "Vegas-style" card games and slot machines, excellent five-star cuisine, and fancy hotel rooms. Somehow, amid all the poverty of former factory towns, gambling has brought a multimillion-dollar surge of revenue into these struggling communities. Casinos have proved to be good for jobs (service and contractor employees), and good for local taxes (Class III gaming facilities are taxed by the state).

Indian gaming profits generated by the casinos go to the various tribes who run them. Much of this new money is used to enhance tribal identity and education. Many Indian nations have built new museums near the casinos that showcase tribal artifacts and culture. The casinos are keeping a delicate balance, however; some of the values espoused by Native American tradition oppose the decadence of gaming, drinking, and nightlife. Yet many of these gaming facilities are at least attempting to help reestablish some of the most important traditions of their American Indian past. And yet these are not facilities in a cultural vacuum. Most of the entertainment is mainstream American. The music is mostly country singers and top-selling pop stars. Some Indian casinos lure people in with pro golf and other sporting events. The featured chefs at these casino restaurants offer cuisines from around the world, mainstays of the "multi-culti" complexion of industrialized modern man. Modern consumerism with an Iroquois twist. In short, it appears as though there is a new era of assimilation by American Indians as they continue to increase their population size, interact economically in modern society, and yet still retain many aspects of their original cultural identity. Thanks to scholarships from casino profits, and the American

Indian College Fund, more young people who claim American Indian ancestry are going to college, which further illustrates the assimilation of Western tradition and American Indian heritage.

Make no mistake. I'm not proposing that we ignore the great inequality and socioeconomic hardship of the American Indian. They live today with many disadvantages, and their history is obviously replete with countless hardships and dispossessions. But I also sense a sort of renaissance of cultural plurality that is apparent throughout the land. As evidence of this, consider that more people visited the Smithsonian's National Museum of the American Indian in 2014 than the National Portrait Gallery in Washington, D.C. More than 1.4 million people viewed Indian art and artifacts at the Smithsonian that year.[13] It's not just gambling that draws people to the ongoing saga of the American Indian; there seems to be a sentimental tie that draws the Smithsonian visitors to seek a richer understanding of Native American culture and its permanence in this country's heritage.

And that underscores the main thrust of this book: In wild species of "nature" as well as in humans, populations have a tendency toward persistence. Once they are established, populations, from microbes to mammals, almost always show signs of expansion, assimilation, and coexistence, commonly in ways that are unexpected. But the thread that sews all of these acts together is the basic tendency of populations to persist. Therefore, if we want to understand the forces that shape our world, we have to understand how populations have clashed, compromised, and persisted throughout the history of life on Earth.

Human populations have the tendency to fight the same battles endlessly. Here in upstate New York we are still squabbling about how land should be used and how we should view land ownership. These days, though, we are arguing about fracking, mineral rights, and—as referenced in the film, *There Will Be Blood*—who is drinking whose milkshake. In the past the Iroquois would have found this fight very familiar. Their culture, and specifically their beliefs about how land should be used, and what it meant to have a surplus, led to many problems with the settlers who squatted in Indian territory. Some Indians claim that those ancient encroachments still have not been resolved.

Populations may value the same things, but they often value them for different reasons. I love the land I live on for its natural beauty, and unlike the gas companies who would love to drill on it for huge profits,

I prefer to keep it free from industrial machinery and pollution. They look at me as a fool for not taking advantage of the money lying beneath my feet. The Iroquois likewise lived in a landscape that from the perspective of the colonists and imperial officials seemed to be all wasted potential. In fact it was far from wasted. Iroquois country was a forested blanket of verdant luxury dotted with lakes, dissected by rivers, and occasionally cleared for villages and their adjacent cornfields. The way of life for these Indians depended on hunting and collecting in the woodlands, fishing in the streams, and harvesting forest products, all activities that today we would call "sustainable" with respect to their population size. This is one of the reasons that their population size was relatively small—only between ten and twenty thousand individuals populated the entire area of upstate New York before the Europeans arrived.

The English couldn't understand why native people allowed so much land to go "unused" for agriculture or grazing, but they never appreciated how well the Iroquois were sustainably using their natural resources. Indian men roamed deep into the woods, while women tended the fields and the houses in the villages. Wild animals and plants formed the core of their subsistence. Since the home ranges of mammals being hunted were so vast, it was necessary to track some of them (deer, for instance) for miles at a time. Others, such as coyotes or wolves, had home ranges that required tens of miles of tracking in a single day. The only way to acquire these animals for meat and fur was to track and kill them in their own habitats, which meant that large parcels of forest had to be left intact, and only small villages on the periphery of these tracts could be maintained (usually on a fishing stream) without disturbing the population balance of game animals.

These two human populations, Indian and European, had completely incompatible worldviews. They both recognized the value of the land, but the imperial worldview was one (shared by French and English alike) that saw land as a resource to exploit, a vehicle for surplus and accumulation of wealth. The native worldview understood land as a means of subsistence, and "wilderness" to them equated with freedom of movement for their wide-ranging lifestyles. Their hunting lands were seen as "wild" by European settlers, but to the American Indians these areas were actively utilized and came with a set of rules, well understood by all tribes, for facilitating the growth of game and other food. American Indians did not see property as a vehicle for accumulating wealth. They conceived of "nature" (a Western term) as a gift that must be respected and cared for lest the giver of the gift stop providing it (as in times of drought, famine, flood, or overexploitation).

In a sense we are confronted today with the same dichotomous worldviews. Given the circumstances that confront us, we as individuals can make very little impact directly on the future of our species. But we can choose to follow one or the other worldview, and try to educate others as to its benefit. On the one hand we have the worldview of the imperialists and Western religions of old, which states essentially: We don't need to care for the planet, it has always been and always will be in the hands of God. Resources were placed here for us to exploit to their fullest, and we are compelled to go forth and multiply. Since we can't feed everyone equally, only the fittest winners in life's ongoing competitive struggle will enjoy the surpluses. The less fit losers will suffer from scarcity. In the end all believers, regardless of their status in life, will be rewarded equally in the bountiful paradise of a limitless afterlife.

On the other hand we can adopt a modern worldview that melds the wisdom from earth science and biology with the ethic of sustainability that American Indians exhibited in their historic past. This will require us to recognize our dependence on the forest and freshwater wetlands that surround our cities, not for surplus or wealth accumulation, but rather for sustenance and health. Using this approach, we can tread much more lightly on other populations, and leave a less indelible footprint on the areas we inhabit. If we agree that preservation of our species is of utmost ethical concern, then it seems there is no better worldview to adopt than this one.

I'm in favor of the modern worldview. It is far from a hands-off approach to natural resources. I value hot water, air-conditioning, travel, and modern life's amenities as much as anyone. I recognize, however, that in order to be a good steward of the planet, I have to consider the populations that are affected by my use of the natural resources I consume. Whether it's energy extraction, ecological conservation, or disease prevention, populations need to be managed. If we can accept this for other species, then why not apply it to our own in the interest of preventing human warfare? In later chapters I will distinguish human warfare from Darwin's "war of nature" metaphor. But this will become clearer after we elucidate the first population wars that started the pattern of coexistence 3.8 billion years ago, back when the planet was young.

2

THE LONG HISTORY OF POPULATION WARS

My sincere love for my fellow man is tempered occasionally by a disdain for the human race. I've interacted with people all over the globe, both as a professional singer in the great capital cities of other nations and as a scientist doing remote wilderness fieldwork in foreign countries. One thing I've learned is that language is less important than attitude if you want to get along with others. I may not speak Portuguese or Japanese, but I know how to order lunch in Rio or hail a cab in Osaka without offending anyone. In other words, even though I can't directly communicate with my foreign friends and colleagues, I can briefly integrate into their cultures. I may not understand their customs, but there are ways in which we commingle and create a functional cultural blend.

This sense of harmony sometimes evaporates when I come back to the United States. Once I'm back home I'm surrounded by people whose views on life, politics, and global priorities often sound completely idiotic to me. No doubt this is because we speak the same language. I'm sure that some of them feel the same way about my beliefs. My relationships with my compatriots run along a spectrum from the great friends with whom I travel on concert tours to the negative, judgmental "trolls" who criticize me for being too old, or too tame, or simply not punk enough to be fronting a punk band. I've learned to ignore most of the ire and recognize it as mere provocation. But other forms of criticism in the past seemed equally out of place.

When I was in high school, Bad Religion rehearsed each day in my mom's garage. Ours was a crowded neighborhood with single-family houses packed five to an acre. Our rehearsal space was not soundproofed initially, so we had strict orders to conclude each day by the time parents arrived home from work (around five in the afternoon). Punk kids, both guys and girls, often came to listen, and while Mom was at work, our house and yard often looked like a nightclub hangout. None of the nearest neighbors complained. A few houses down, however, lived a family of fundamentalist Christians who viewed my family with contempt. Although they never directly criticized us, they were the least friendly family on the entire block. On one occasion, shortly after the band was formed, they called the police to try to stop our rehearsing. After some questioning, the cops were satisfied with our five o'clock "cutoff" time, they implored us to install some sound baffling (which we promptly did), and they left us to rehearse. One of them told us who had called in the complaint, and it was indeed the unfriendly family a few houses down.

I'm sure that, like many fundamentalists, this family simply couldn't be open to the fact that ethical kids can come from many different types of family situations. In our case music and punk lifestyle formed our daily after-school rituals. They probably thought that we were ruining the fabric of their society with our "devil music" and threatening their children's morals with our sinful behavior. They lumped me into a population of heathens, people they had only read about, and assumed that I was therefore at odds with their way of life. In actuality we were of the same population—modern America—even though our views differed on certain topics.

Today, whenever I talk to Christians, as a teacher of evolution I am conscientious about not assuming too much about their views on science. I have to be very careful in such conversations. I have no problem openly discussing many of evolution's implications, such as why I believe life has no ultimate meaning, but this idea is a conversation killer with devout believers. Even though we can live on the same street, there's no denying that our worldviews have inherent incompatibilities. While a lot of people are content to judge silently, I'm more interested in discussing the merits of both views with the intention of agreeing rationally on a conclusion. When we make some topics taboo we end up with an entrenched and irrevocably polarized society. Whereas when we broach difficult subjects with openness and mutual respect we at least have a shot at ending up with a happier and more balanced citizenry.

I have had to hold my tongue in the interest of maintaining good relationships with my neighbors, most of whom probably believe in God. I never got into a fistfight over religion, and I avoid arguments about spiritual beliefs or political ideologies that I know aren't going to go anywhere helpful. I intentionally avoid confrontational topics with the people whom I engage because in most cases we are trying to resolve something productive.

This is especially true with my colleagues. I argued constantly with the other members of Bad Religion during the first half of the band's life. Usually it was over the logistics of touring, something simple such as: Should we travel by day or cruise to the next city overnight on the tour bus? or Should we play the same songs each night or change the set list occasionally? One day we were trudging our way through the same never-ending touring disagreement when I snapped and said, "Listen! We need to compromise, which means that none of us is going to get his way on this issue." I don't even remember the details of the disagreement, but the point was clear: When the two sides of an argument are incompatible, we have to give up something of our self-interests for the sake of the band. In other words putting the band's interests ahead of our own was ultimately the best thing for all of us. *Compromise means that nobody gets his way entirely.* But in order for it to be productive, we all agreed on a trade-off. Something beyond individual, selfish needs was driving the whole enterprise forward.

That moment and its positive repercussions gave me firsthand evidence of the benefits of setting aside self-interest for the common good. Our band's subsequent tour was more successful than it would have been if we hadn't compromised. Everyone involved—band members, touring agents, and fans—benefited from our ability to put aside our short-term feelings and think about the long-term functionality of our enterprise.

The concept of compromise for the sake of function that we stumbled upon in that argument is at the heart of human coexistence. I believe, furthermore, that although compromise is a concept in conscious human relations, it has parallels in the unconscious unfolding of human history and in the evolution of other species as well. For me compromise has become a metaphor for one of the ways population wars are played out.

No organism in the natural world exists in an idyllic, unperturbed vacuum. Animals, plants, bacteria, fungi, and protists are everywhere, interacting with other organisms on many fronts as they live out their moment-by-moment existence. Life would cease altogether without this elaborate web of symbiosis, yet for the most part we go through our lives completely unaware of the drama unfolding around us. Scientists prefer to use the word "equilibrium" rather than "compromise" when describing the interactions of populations. But to anthropomorphize it, the concept is essentially the same: No population exists without the cooperation of other populations. More universally, then, we can say that compromise means that no groups get their short-term goals fully realized[14] because they participate in a web of coexistence. For humans we often consider the social web as primary—who gets along with whom—but there is an equally vital biological web of organisms and cells to consider for each human being, as we will see. For nonhuman populations we consider the ecological web—ecosystems and their associated populations—as mitigating their relationships.

One of the surprising fallouts from the last two hundred years of scientific progress is that most people now know who Charles Darwin is, and what the main thrusts of his arguments were, but they rarely understand them. Darwinism has been misunderstood and often distorted to justify cruel sociological philosophies. This is especially true in the popularity of the idea that the "best" species or individuals will "win" some sort of competitive race called "life." While it is true that the individuals with the most useful adaptive traits will most likely survive, and have the most offspring, there is no value judgment involved. This individual is not the "best"; she is simply the one whose historical and biological circumstances gave her the greatest odds of survival and reproduction at one specific moment in time.

One of my biggest pet peeves is when people read about a stupid accident—often one that has resulted in the death of an intoxicated college student—and shake their heads, saying, "Darwin in action." We've all done stupid things in the past. The fact that we were lucky enough to drive home without getting into a wreck, or to avoid picking a fight with the wrong person, is not evidence of genetic superiority. Only a callous person exercising a gross misapprehension of science would pass judgment on someone less fortunate than himself. Now imagine this judgment applied to a whole population, be it one that belongs to a different belief system, race, or nationality than our own. There's no justification for using Darwinian evolution to pass judgment on others. But evolutionary science is the best tool for understanding the ways that population wars play out.

There are four traditional categories of species interactions that fall under the heading of symbiosis: parasitism, predation, mutualism, and commensalism. The first two of these are antagonistic in the short term. Parasitism, the familiar situation where one species infests, infects, or inhabits the bodies of another species, and predation, where one species uses another as its source of food, usually make for dramatic storytelling. Familiar examples of these antagonistic types of population war include lice inhabiting the hair of schoolchildren (parasitism) or Canadian lynx chasing and killing snowshoe hares (predation).

The first of the two less antagonistic relationships between populations is mutualism, where two species derive benefits from and provide benefit to each other. Both species give up something in order to accommodate the other. This can be seen, for instance, in many species of ants (leaf-cutters) in the tropics. Huge colonies composed of thousands of individuals create elaborate underground caverns as part of their communal nests. Inside these subterranean burrows are gardens where the ants deposit leaf fragments they have harvested from the surrounding forest. Fungi grow on the leaf fragments in the gardens, and the ants make a meal of their vegetative parts. The fungi are not killed in the process; they just keep producing the vegetative portion as the ants feed over time. The ants benefit from having a convenient and predictable reserve of food growing in their nests, but the elaborate burrow construction and huge effort it takes to cut leaves and cultivate the gardens is a cost they must endure in the relationship. The fungi benefit from having a safe haven protected from other scavengers and parasitic bacteria of the forest floor, but have only restricted opportunities (gardens) to grow, thereby decreasing their opportunities to feed on and expand to other areas of the forest.

Commensalism is the second type of less antagonistic relationship. A commensal species is one that derives benefits from another species but doesn't inflict any costs on the other. An example of this is the remora, a strange fish with a specialized sucking organ on the top of its head that allows it to attach to another species (usually a shark). Remoras "hitch a ride" on their hosts without actually providing any help in locomotion, or any benefits to speak of. There are no known negative effects caused by their freeloading. Usually the commensal individual is so small relative to the mass of the host that no significant friction or drag is caused to hinder transport. Another example of commensalism is the presence of possums, raccoons, suburban coyotes, or even bears that regularly eat the trash we throw away. These species opportunistically feed on our refuse, and there seems to be little or no cost to us associated with their commensal activity.

One chore I enjoy is cutting the grass. It just so happens that our "grass" is an alfalfa field nearly four acres in size. Every spring and summer I resign myself to a few days of driving my old tractor, mowing the tall grass. It's hot, loud, and dusty work, and I'm pretty tired of the process after a few hours, even though the end result (hay) is a good thing. However, my mowing has a bloody side effect. The land is full of different small animal species, and as the blades cut the alfalfa they inadvertently slice scores of creatures unable to escape from under the mowing deck. Most are insects, but occasionally a field mouse or a vole runs the wrong way and can't escape in time. This massacre is a boon to the crows that live in our woods, and they quickly swoop in to feast on the corpses. The birds are loud and obnoxious, and sometimes I have to swerve to miss them. I'd rather they go somewhere else. But I also know that they enjoy having me cut the grass, sometimes running over what will become their next meal. The crows have a commensal relationship with me, just like all the suburban commensals I mentioned above. I tolerate the crows, but they benefit more from the relationship than I do.

It's easy to see how someone might construe the antagonistic relationship between a predator and its prey, or a parasite and its host, as that of a victor (the predator or parasite) and the vanquished (the prey or host).

It's important to remember that if the host or prey is exterminated, then the population of parasites or predators can't last for long. The ways in which species interact is a more complex process than most people realize. If you haven't studied the subject it's easy to misinterpret Darwin and imagine that the natural world is in a constant state of high-stakes conflict. At the very least you might imagine a world where one population regularly dominates another into extinction, either by eating it or taking away its vitality. However, extinction in nature does not often result from the direct actions of other populations. Instead there is a spectrum of symbiotic relationships.

Even the most antagonistic relationship results in an equilibrium over the long term, which for our purposes can be viewed as a compromise between species. This is most glaringly illustrated when an incompatible pair of species comes together in a predator-prey relationship. The classic studies of predators and prey in the wild show oscillations of size for both populations. When predators are abundant, they reduce the prey population. Lower numbers of prey mean a limit on the predator's reproduction rates, and their population size soon falls. This in turn makes for fewer predators, so the prey population increases once again. Over time these fluctuations reveal a stable equilibrium between predators and prey.[15]

An often-cited example is the wolf and mountain lion populations and their prey, the mule deer, in the Kaibab Plateau of Arizona and Utah. It is clear that one species, the deer, is the food, and the others are the eaters. There is no immediate short-term compromise. The mule deer are the prey, and until the recent past they were a valuable food source for predatory mountain lions and wolves. While this might have been unfortunate for the individual mule deer, the relationship between the three species worked well. They were linked together in an ongoing cycle of predators and prey, and their population sizes reached a stable equilibrium. However, during the nineteenth century, the deer, wolf, and lion came into contact with a new kind of predator—the cowboy. Ranchers brought in herds of sheep and cattle and wanted them protected. Mountain lions and wolves were no match for the Winchester 1873, and within a few decades the dominant carnivores of the Kaibab Plateau were near extinction.

This kind of overhunting causes problems. As we now know, you can't get rid of one species and assume that everything else will stay the same. Populations grow by their own intrinsic rates of increase until limits are encountered. The extermination of the wolves and mountain lions removed these limits for the mule deer, and their population exploded. The deer, which numbered four thousand in 1906, were estimated to have increased to as many as one hundred thousand in the first thirty years of the twentieth century. It appeared as though the mule deer ended up being victorious in the population war, even though they were the ones who had been hunted by the predators for most of the previous century and before. Furthermore, their "victory" wasn't even self-made. It was the removal of predators by humans that sealed their fate.

What happened next illustrates one of the central themes in evolution, and something that I urge every reader to consider when thinking about our own future as a species. Population dynamics cut across generations, and sometimes vast expanses of time. Something that appears to be positive for a species in the short term can turn out to be a huge problem for it in the long term. Shortly after reaching their peak in population size, the mule deer population experienced a 60 percent loss. Nearly sixty thousand deer starved in only a few short years. This is not the story of a victorious population. The deer population experienced a literal boom and bust, and arguably the individuals experienced more suffering through starvation than did their ancestors through being occasionally preyed upon.

Similar stories are still playing out all over the world: In 1859 the first fleet to Australia brought rabbits as a food animal. With no natural predators, they reproduced and grew as a population that became "out of control." An exceptionally bad drought in 1988 meant that millions of rabbits starved, but that ultimately only dented the population. Today hundreds of millions of rabbits cause serious crop damage and habitat degradation for native species.

With no natural checks on its population size, any species will become invasive and overpopulated. The Chinese continue to pull sharks out of the oceans through industrial fishing. We may not like sharks, but oceans without sharks become overrun by other species, such as squid and jellyfish. This is a serious problem that could lead to a great biological crisis in our oceans. Disruption of a single species often has an impact on all others due to the ecological web of coexistence in any ecosystem.

What appeared to be a short-term gain for the mule deer—removal of their nemesis predators—ended up becoming a long-term tragedy involving mass starvation of their herd and disruption of their habitat by herds of livestock. Finally, however, the population leveled off to between ten and fifteen thousand individuals, where it still stands today. Their old terrain has changed, and the mule deer will continue to have to deal with new competitors; in this case cattle, horses, and other ranch livestock.

These relative newcomers can be considered "invasive species." Like most invasive species they have a rapacious thirst for water and resources, quickly sucking up more than their "fair share," and leaving the native species vulnerable to desiccation and therefore in a very precarious position. Whenever we humans modify a habitat for our own needs they are considered "disturbed." We are only just starting to realize how much harm we can do by opening up roads and stringing barbed wire through prairies, clear-cutting timber in forests, and removing alpha predators from the oceans or mountains. Yet, even after hundreds of years of thoughtless behavior, it is still possible to salvage seemingly ruined landscapes and environments. While it's the natural tendency of all populations to reach some sort of equilibrium with the others in the biosphere—a process that occurs over many generations—in disturbed areas it is now a matter of careful stewardship to restore a sustainable balance. The mule deer now have to contend with an impassioned human drama over reintroducing predators, which is playing out across the West. I believe that reintroducing wolves to the West is a good thing, though it has caused a huge amount of anger and frustration among ranchers and hunting guides who believe that the wolves are sport killing livestock and wild herds. There is an ongoing debate about how the wolf fits into the modern mosaic of wilderness and ranch and town life that makes up America's Western landscape.

Right now you might be thinking, Why should I care about long-term population dynamics and compromises when I have to take care of myself and my family in the short term? Why do I have to care about evolution? The answer is you don't *have* to; but if you want to be a steward of the environment, even a steward of your own backyard, community park, or the health of your own body and that of your family, you need to care deeply about it. Any scientist, especially one working in ecology or climate-related specialties, knows that persuading people that they should care (and that their attitude can make a difference) is one of the great challenges facing our planet. It's one thing to know that the world's equilibrium is disturbed, but doing something about it is another. We have to reconcile the knowledge that our population is out of control with the desire to take action and do something. One of the things that will help us to do this is developing a sense of empathy.

If we accept that our sense of emotional well-being is dependent on having a sense of purpose during our lifetime, then it is crucially important to determine our place and role in a "bigger picture." The way we define that idea has changed radically over the last hundred years. We have to remember that for literally thousands of years we defined what was good and meaningful by a traditional religious narrative. This narrative is slowly falling by the wayside (though it isn't going without a fight) and is no longer being used to direct national or global policies. We are depending more and more on rationalism, scientific innovation, and information sharing than ever before. This is a massive transition; and we are still defining what a secular "bigger picture" looks like, but clearly most modern citizens have access to the Internet and can see that others in faraway places share many commonalities. This might rank as the most important social awakening in human history. Part of this renaissance includes a fresh ethical approach that values community over cultural pride—in other words a worldview that is less about being part of a regional, ethnic, or sectarian population and more about being part of one global population.

If your only concern is for a better afterlife, then you needn't incorporate any scientific knowledge or modern information technology (such as the Internet or mass media) into your daily life. The paths to salvation have been clearly laid out for centuries by religion. But if you are willing to expand your worldview and try to see yourself as a global citizen, then you need an awareness of how your basic needs are related to the environment.

Once we understand that compromise is part of the deal, and that no one gets his or her way all the time, we are more likely to make decisions that will not only benefit us and our families but also benefit the bigger picture. This might seem impossible or contradictory, but you've been doing it all along, perhaps unknowingly. Separating out the recycling isn't a particularly enjoyable activity.[16] It's smelly and messy, and frankly I'd rather not have to do it. There is no discernible plus for me to rake through the stale food containers, but I do it, as do most of us. The sorter is taking an extra step to do something that benefits the larger social group far more than it benefits the person doing it. Recycling your trash is a pain in the ass, but it's better for the environment than dumping it all in the garbage. It's also part of a bigger social correction.

Even though recycling may sometimes seem like a largely symbolic gesture, it has actually become a part of our cultural narrative since the 1970s and is a massive shift from the carelessness of the early and mid-twentieth century. When I was a kid we didn't care too much about littering or pollution or species eradication. I've heard "It's good to reuse things instead of dumping them in a landfill" since I was a young boy, but it always sounded like good old-fashioned Midwestern pragmatism.

Today recycling is a philosophy, and a global industry, that benefits the environment on which we depend.

What we think and what we do are tightly interwoven. For instance, American culture (like most others) has always glorified war; this acceptance and even approval of armed conflict has led to violence that in turn justifies further global conflicts. But it doesn't have to be this way. We can redefine war in the context of what we've been discussing in this chapter—as an inevitable result of population interactions through time. In this view, the vast majority of human population wars have resulted in a blending of culture and biology that over the long term become recognizable as human assimilation. Seen on a spectrum, it can be said that the end result of war is neither complete harmony nor total incompatibility. Only very rarely do wars end up at either of these extremes.[17]

Our history is full of cultural assimilations that have been told as war narratives. The French and Indian War (1754–63), for example, was less of a "good-guys-against-the-bad-guys" story than it was the culmination of an assimilation process that played out over hundreds of years after European contact with American Indians. I live in the heart of historic Iroquois territory; I can drive a few minutes to where the skirmishes and battles took place. It's hard to reconcile the brutality described in historic journals with the peaceful glens and hills that surround the Finger Lakes. The American Indians' lives were changed forever due to this interaction, but the lives of the Europeans and immigrants were as well. The end result of these battles was assimilation. Both French and "Indian" populations are still with us today, and the descendants of their early contact now identify as a distinct ethnicity (métis). They relate to one another in generally respectful ways.

The only logical thing about human warfare is the feelings it evokes once it is over. It seems to be almost universally reported, in literature and personal testimony by those who have seen combat, that after a war those who have gone through it consider it to have been traumatic. It's perfectly understandable that they would not wish such a harrowing experience on anyone. Even though most American citizens have not gone to war, we feel sympathy with veterans, past and present. How we act on those feelings is determined by cultural narrative. We believe that "bad guys" caused this suffering of our veterans. So what are we to make of such beliefs? Retribution in order to vanquish the enemy is a logical conclusion. If war remains a story in our collective experience that pits good guys against bad, we are destined to continue the cycle of destruction.

Most students learn the basic why, who, and how of warfare in history classes. However, in order to understand conflict you have to look at the natural world as well. As it turns out, the way we have been taught to understand wars in human history is not at all consistent with what nature tells us about the possible kinds of organismic interactions. War, when viewed through the lens of populations coming into contact with one another, is an elemental component of humankind, and of all non-human life as well. This is not a moral judgment or a philosophical statement, but rather the logical conclusion from studying a process that has been going on for billions of years. Populations of cells as well as individuals of all species increase due to inherent mathematical properties. Eventually they come into contact with other populations, and Darwin's "war of nature" becomes inevitable. The growth of populations results in clashes that continually alter the course of evolution. Until populations reach their equilibria, suffering and hardship become part of the equation. Unlike other species, however, we humans have the ability to limit our future suffering by learning to avoid the pitfalls of past narratives and act on new ideas.

Suppose we decided to develop a modern ethic for living our lives in a way that promoted the longevity of *Homo sapiens.* I would attempt to build one based on the following evidence from nature. We have enough data from paleontology, geology, and biology to conclude that 99.99 percent of all species that have ever lived on the planet are now extinct. Our own species, *Homo sapiens,* is a relative newcomer; we evolved only within the last two hundred thousand years, yet the natural history of our planet is 3.5 billion years old. Ten thousand years ago, when agriculture began to flourish, our numbers and distribution exploded. We now have the machines and weapons to inflict widespread death on our own kind, and to poison irreparably the habitats upon which we depend. But we also have the tools to avoid extinction. We have come to a lofty social plateau: Unlike any other species, we have the ability to make sophisticated predictions about the future based on knowledge of the past. We know how things tend to turn out for species that can't adapt as the world changes around them. Striving to be in the 0.01 percent of species that don't go extinct seems a worthy pursuit. It seems to me, in fact, more of a moral and ethical imperative. Shouldn't our policies and goals be geared toward the longevity of our own species? What better moral position could one take?

Let's think back to what the world looked like when there was no human strife, conflict, or conventional warfare. There was a time when no populations were found, neither human nor other species on the face of the planet. The early Earth was very different than the one we live on today. It had no life and was simply a young planet with a very inhospitable environment.

The earliest life on the planet sparked into being at least 3.8 billion years ago.[18] The sun was about 25 percent less intense at that time, and the blanket of gases we call the atmosphere was not well developed enough to keep things warm at night. Consequently any heat generated from the molten core of our young planet dissipated into the night sky. Still, localized microhabitats with high salinity and geothermal heat were common. In these places the first life-forms probably evolved.

There were oceans. We know this because geologists have discovered rocks, originally made of detrital sediments, that date to 3.8 billion years ago (bya). There can't be sedimentary rocks unless there were sedimentary basins (that is, oceans) to receive the sediments. Even though the exact chemical composition of seawater at that time is not precisely known, the sediments had to have come from somewhere, and then as now, erosional debris indicates that a landmass was not far away. The chemistry of the oldest rocks indicates that a blanket of CO_2 seems to have been building up to form some kind of rudimentary atmosphere. So, even though it was still very cold, there was enough of a gaseous blanket around Earth to keep the planet unfrozen and make it favorable for the synthesis of biological molecules (biomolecules) in the primitive oceans.

In a classic 1952 experiment, Stanley Miller, working with his graduate adviser, Harold Urey, synthesized biomolecules in the laboratory. He simulated the early ocean in a test tube and added inorganic chemicals like the ones that might erode from landmasses and spill into the nearshore environment. Assuming that the atmosphere would produce lightning as ours does today, Miller added a spark to the gas over their primitive soup of ocean water and inorganic chemicals. The spark synthesized organic molecules from inorganic molecules. In fact, with simulated lightning as an energy source, a hyperabundance of organic building blocks (amino acids) was synthesized and condensed in their laboratory apparatuses. All life as we know it is comprised of just twenty different amino acids. But Miller and Urey's experiment produced an even greater number of amino acids than are found in life forms today.

It may have taken hundreds of millions of years, but after the building blocks were created, other fortuitous events resulted in the formation of membranes, and eventually in the assembly of those membranes into cells. But before cells evolved, there were simply organic molecules floating about in a global oceanic oxygen-free environment. These microscopic first steps of life all occurred in the absence of any kind of reproducing species, and therefore populations didn't exist. In fact the next step was the origin of the first species, which must have found the world a lonely place.

The oldest fossils reveal that at least one species of microscopic organism had evolved by roughly 3.5 bya. These filamentous, colonial organisms are known as cyanobacteria. And they are still with us today, living in very inhospitable places—perhaps vestiges of their preferred habitats on early Earth. Then as now cyanobacteria used a chemical process called photosynthesis to transform atmospheric CO_2 into building-block biomolecules and excrete oxygen (O_2) as a by-product. Sunlight is needed to drive photosynthesis. With a virtually unlimited source of atmospheric CO_2 (from volcanoes), the first organisms on the planet had no competitors. As long as the oceans remained unfrozen, the early cyanobacteria could proliferate in the upper parts of the ocean where sunlight was available. But the proliferation of these photosynthetic bacteria came with a cost. Remember that these microscopic life forms excrete oxygen as a waste product. Like most organisms, these bacteria found that living in one's own excrement can be unhealthy. In fact many bacteria die in the presence of oxygen. In order to tell this story at a basic level, it is necessary to point out that this is a drastic oversimplification. Most of the oxygen produced in the earliest stages of photosynthetic organismic evolution must have gone toward oxidizing sediments. It is hard to speculate on the number of years that it took to totally saturate all the early sediments on the planet, but it was only after that equilibrium was breached that oxygen became liberated into the atmosphere. The atmosphere began to fill with a molecule that is toxic to the very ones responsible for its liberation. So began the world's first population crisis.

Life has always had a cost. Even the earliest and simplest life forms profoundly affected the environment in ways that were detrimental to their well-being. It's the nature of populations, through their intrinsic rates of increase, to cause problems for themselves.

The organisms that produced the first oxygenated atmosphere were faced with a problem. Their own success had created an environment that was now toxic for them. As a result their population most likely crashed. In order to survive they had to evolve some sort of mechanism to withstand the trauma of an oxygenated environment. Slowly, over the course of countless generations, a thick capsule-like coating evolved that completely covered every cyanobacterial cell. This thick coating[19] insulated the cell and its genetic contents (DNA) from the harsh environment. Excess oxygen went to oxidize or "rust" sediments. These can be seen in the stratigraphic record as "banded iron formations" (BIFs) that date from roughly 3.8 billion years ago. By roughly 1.8 billion years ago Earth had a biosphere and ecosystems full of diverse anaerobic bacteria and algae, all producing toxic O_2 as waste. Over the course of nearly two billion years numerous species evolved in multiple lineages. Some were anaerobic solitary cells and functioned in the absence of oxygen. Other lineages had scores of species photosynthesizing and possibly competing with one another for access to sunlight. Rusting of sediments tapered off by around 1.8 billion years ago (BIFs of that age and younger become rare), indicating that a balance was reached between the production of oxygen by populations of photosynthesizing organisms and consumers of oxygen.

The complexity of microbial interactions became established at this time. All of these planktonic populations floating around in the primordial ocean came into contact with one another. Some populations were likely striving for more sunlight, others migrating to areas of lower oxygen concentration in order to regulate their exposure to its toxicity, maybe others were aerobic and moved toward populations that produced abundant oxygen, creating some sort of primitive commensalism or mutualism. The limited resources of oxygen and sunlight provided enough impetus to cause the earliest population wars on our planet. From them the stage was set for new groups to evolve. At that point in time, oxygen and its toxic effects became a mainstay of Earth's ecosystems.

This was a quietly profound moment in natural history. Then as now, individuals were thrust into a cycle of proliferation, symbiosis, and hardship, all by their own unconscious actions. They were, as we are today, propelled along this path by the unseen chemical changes in their cells that, without their knowledge, compelled them to multiply and reach equilibriums in a world full of toxins and sometimes overrun by individuals from other populations.

Humans are the first species in 3.5 billion years of evolution to have the awareness of population wars. Unlike any other species that has ever lived, we have the ability to consciously improve the odds of our long-term existence. Maybe we can also achieve the longevity of some other species known as living fossils.

I take my family out for sushi regularly. Invariably, these restaurants are decorated with plants of the genus *Equisetum*, also known as horsetails or snake grass. The botanist in me can't help pointing out that we are rubbing elbows with one of the most primitive plants in the world. It has thrived since the Paleozoic. Horsetails are found as fossils from Kentucky that date to the Carboniferous period, more than 300 million years old! This means that somehow, despite great upheavals of Earth's climate and shifting continents, and literally millions of extinctions of other species, this group has maintained an unbroken line of ancestry for hundreds of millions of years. For this reason *Equisetum* is considered a "living fossil"—an organism known to have survived through vast expanses of deep evolutionary time. Horsetails prefer to grow in wet, lowland, soggy settings, far from rushing water in the quiet overbank mud found throughout continental habitats. We know that these depositional settings have been common throughout the last 540 million years because they are preserved as mudstone in the rock record. Given the ubiquity of their habitat choice—along the major river systems on all continents—and the low incidence of disturbance from herbivores (not many animals prefer the muddy overbank areas), perhaps the horsetail found a fortuitous combination of factors that aided its long-term survival.

Other living fossils can be found in the United States, each one in a similarly unique and persistent environmental niche. On the Atlantic coast lives a species of arthropod called the horseshoe crab (genus *Limulus*). They stay relatively near shore in waters less than one hundred feet deep; once a year huge numbers of them come ashore to mate in the sandy shallows and, as an unintended side effect, scare and fascinate small children with their alien appearance. These large crablike animals (known scientifically as chelicerate arthropods) are dark brown and about the size of a halved volleyball. A more accurate description might be, at least according to some of my friends, an "extremely large aquatic spider with hard shell, long tail, and generally alarming appearance." Whatever charm they have is certainly not in their looks, but you have to admire their incredible tenacity and survival skills. They are almost unimaginably older than we are, but seemingly have never undergone a significant anatomical change. Horseshoe crabs have been discovered in Upper Ordovician sedimentary rocks dating to 445 million years ago. Earth has gone about its population wars and extinctions and asteroid bombardments, glaciations and hothouse phases, but the horseshoe crabs just continue their seasonal mating efforts in the intertidal sands of the seashore.

Living fossils such as these have no awareness of their long-term success. They have found a place in the biospheric milieu, a niche that suits them perfectly, and is persistent through long spans of geologic time. Such species seem to have reached a lasting equilibrium with the biosphere that is as strong today as it was in the past. Their equilibrium is robust enough to survive pollution, human population encroachment, climate change, and a constant creeping of the substrate from plate tectonics. So, if they can survive, how about us? Can our species forge a similar lasting compromise with one another and our environment? We have the opportunity to be one of those rare exceptions to the 99.99 percent extinction rule, and instead be the .01 percent that avoids extinction. But this requires an across-the-board commitment to understanding the historical circumstances that brought us to this point in time. For this we have to study our world, assess what is happening to it, and make choices about how to treat our environment and one another.

It may sound peculiar coming from an old punk rocker, but I strongly believe that governmental policies are the only viable way to administer our long-term success as a species. I guess you could say that my attitude of "fuck the government" is still intact. But it's more a criticism of lousy government than a statement of nihilism. The truth is, when it comes to environmental protection, the government is the best way to enact a new social awareness by establishing laws by which industries have to abide.

If I get some free time in the summer, I like to take my family hiking in the Sierra Nevada of California, or some other getaway near Los Angeles. If I'm lucky we get to make a multiday trip. Sometimes our annual hike is limited to a day trip to the Channel Islands, the least-visited

National Park in the continental United States. An excursion to the islands means getting up early to catch the one daily ferry. Once we make the crossing we spend the day observing endemic species like the Channel Island fox, and reveling in a stunning wilderness. Some relatively untouched land surrounds Los Angeles, despite its reputation as a polluted parking lot. And we are lucky to have a local government there that, finally, takes protecting it semi-seriously. It's taken a while, and we've lost some land that by rights should have been preserved as wildlife habitat and public lands.

Overzealous "developers" tore up the Santa Monica Mountains in Los Angeles and Ventura Counties, ruining pristine wilderness and replacing it with tract neighborhoods and multifamily dwellings. Why wouldn't they? This area includes Malibu and its mountainous backcountry, home to rock stars and Hollywood's rich and famous. The rugged mountains, full of twisty roads and some absolutely gaudy estates, drop precipitously to the beaches of the Pacific; many of which are completely inaccessible to the general public because they are bordered by other gaudy houses. This is all the more galling because the beaches are public land. The wealthy elite of LA and San Francisco spends a small fortune in court trying to repel beachgoers from exercising their legal rights to enjoy the sun and surf.

Scientists had known for years that the coastal chaparral plant community that drapes the mountains and extends to the coastline is fragile and rare. This kind of flora community is only found in small pockets of land that experience a Mediterranean climate. These habitats are typified by short, shrubby vegetation with curled leaves (reducing the drying effect from direct sunlight), occasional hardwoods such as "live" oaks, whose leaves have waxy coatings (reducing evaporative water loss), and a constant threat of fire from the unusually dry conditions. This all sounds inhospitable to many, but in fact the Santa Monica Mountains are known as a hot spot of biological diversity with hundreds of endemic plant and animal species that occur nowhere else on the planet—all surrounded by one of the world's densest human population sprawls.

In order to protect the native vegetation and animal habitat, new laws had to be drafted. All building permits have to pass through the California Coastal Commission, and each proposal to build is scrutinized by a committee of scientists. Even well-to-do celebrities get rejected on the grounds of environmental impact. A famous case involved U2 guitarist

"The Edge," who purchased 156 acres of wild chaparral but wanted to build five mansions on it. Needless to say there was going to be a significant disruption of the fragile habitat, and his building plans were rejected. The executive director of the Coastal Commission called it "one of the three worst projects that I've seen in terms of environmental devastation." Their refusal to rubber-stamp projects is proof that local government can indeed protect the habitats and species of ecologically fragile areas.

You don't have to be involved in local government to know that its policies play a major role in environmental health. Los Angeles used to be known for smog. It was supposed to be a land of beaches, mountains, and sunshine, but by 1976, the year my mom, brother, and I moved there, a deep brownish orange cloud of pollution hung over the skyline ridges of the city every day. There were so many automobiles and the climate was so dry that exhaust from the cars and industrial plants was accumulating and not dissipating. This concentration of gaseous and particulate airborne waste became a symbol of city life. Aside from the health risks associated with this toxic cloud—asthma and other breathing disorders—it also made LA the butt of many jokes and social commentaries. Why would anyone want to live in a basin of smog?

The tide began to turn thanks to governmental policy and forced compliance with laws that targeted the polluting industries. Catalytic converters became mandatory in 1975 and unleaded gasoline (which is the only kind that works with catalytic converters) began to replace leaded fuel. Today the skies are much clearer in Southern California as well as in other urban centers throughout the world thanks to the standards that were created by laws to protect the environment. More recent reports indicate that LA is not out of the woods yet. Although the skies are clearer than they were in the 1970s, there are some unregulated industries and sectors of society, such as boats and industrial equipment, that contribute dangerous levels of ozone into the Southern California basins. But better engines can be built to stricter environmental standards. What is needed are the laws to create cleaner air.

Governmental policies have also contributed to the protection and longevity of species. The brown pelican is a classic Southern California example. By 1970 the pelican was on the endangered species list. Next to the pelican was the most symbolically important animal in America, the bald eagle. Both of these species were experiencing a drastic reduction in numbers due to the indiscriminate use of a manufactured toxin called DDT.

DDT was used as a pesticide for many years in the twentieth century. It increased profits for some, such as those who manufactured the chemical, and for some industrial farmers by increasing their production of certain crops, but it also caused misery to many. DDT has a nasty feature: Once it's applied, it breaks down very slowly. That means it stays around in the soil and gets on the animals and plants of nonsusceptible species. The DDT then in turn gets eaten by other species. It turns out that DDT caused particularly serious problems in many bird species. The chemical affected eggshells and made them very weak and prone to breakage. As a result numerous species, including the bald eagle and brown pelican, were experiencing very low reproductive rates: None of the fertilized eggs could make it to hatching.

The governmental policy to ban DDT was hard fought. Industries and agriculturalists didn't care about the environmental damage; they just wanted to focus on profits. It didn't matter to them that once they applied DDT, nature couldn't get rid of it. It just continued to build up in the environment where it was sprayed year after year. Even today, fifty years after the ban on its use in agriculture, it still is found in places as disparate as the blood serums of California sea lions and human beings in Midwest farming communities. DDT has therefore passed throughout the complex food chains of so many species that it has gotten concentrated in nearly all animals that eat other animals. The fact that it shows up in human blood samples is particularly worrisome. DDT has been linked to the increase in diabetes as well as to many kinds of cancer.

Government environmental policy is our best hope for the maintenance of the biosphere. Nature has done it without our involvement over the course of countless millennia before we evolved. But it's likely that our species, at its current rate of growth and consumption of resources, doesn't have the luxury of time. We do have the ability, however, to consciously choose to effect that balance faster than it was ever achieved in earth history. Conscious adherence to policy is one of the tools in our quest to become global stewards. The ban on DDT saved some species from the brink of extinction. The bald eagle and brown pelican have been taken off the endangered species list. Their populations will begin to stabilize now that we have brokered a grand compromise between them and the farmers who have other pest-fighting materials at their disposal. Through smart policies such as these, we can engineer a more balanced, sustainable future for ourselves.

By following this story of compromise and cooperation, we begin to realize that trying to assign values of "good" and "bad" to the various conflicts of the past is pointless. Rather than try to figure out who is the worst transgressor, we need to understand what brought these groups into contact, and see if by understanding the chain of events and their causes (be they biological or historical) we can find some empathetic middle ground that allows us to listen to each other, and respond constructively rather than destructively. This is a long-term proposition; it requires thinking not just about our needs but the needs of people ten or twenty or two hundred generations down the line. If we are committed to having the longevity of a living fossil, then we need to exploit our unique ability to review our past, assess our present, and plan for our future. Only by doing this can we hope to reach a lasting equilibrium with each other and with the other species on the planet.

3

THE MEANING OF COEXISTENCE

Human populations can hardly be characterized as currently living in any kind of equilibrium. Rodney King, the reluctant face of the 1994 Los Angeles riots, famously asked, "Can't we all just get along?" The answer to his question is, "Not really." Without a unifying principle or ethic, the human race will continue to be fractured into subpopulations that have only their own self-interests to guide them. So far no human group has developed and passed along the cultural narrative that is necessary for peaceful coexistence.

Unfortunately most groups of humans lack empathy for "others," and without it there is no common goal. The idea of working together for a sustainable future as an ethical imperative is a new concept; it may just take a long time for humans to figure out how to avoid conflict and cooperate instead.

Once we accept the long history of population wars we have to conclude that they are inevitable, and this means that we cannot get along unconsciously. Getting along—like most meaningful human endeavors—takes effort. It will require education, governmental policy, heightened awareness, and in some cases enforcement to do so. Look at something as simple as plastic shopping bags; we all know they are "bad" for the environment, but for the most part we stop using them only once a local ordinance bans them from our shops. These kinds of seemingly petty regulations—no matter how much they may gall our libertarian friends—are necessary. They are a small part of a bigger shift that will tweak the norms of behavior to benefit the environment and enact a grand compromise for the good of our species.

So how do we "all get along" as a species? It can't be as hard as some would make it seem; the natural world is full of species that are able to do it despite their seemingly incompatible needs. The answer is for mankind to become a race of enlightened citizens among the community of other species with whom we coexist. The first step in our collective raising of consciousness is to look at how coexistence has been achieved in the natural world, and in the unconscious unfolding of human history. Throughout history we have been involved in grand compromises without even being aware of them.

On a recent drive along the country roads near my property in upstate New York, I passed a mangled piece of roadkill. Since I live out in the countryside I usually spot one or two deer carcasses a day. These deaths are unfortunate but inevitable. The deer like to graze along the grassy shoulders of the highway; as they eat they get spooked by traffic and dart into the road, where they are hit by oncoming vehicles. Their remains litter the sides of the road, and drivers rarely—if ever—stop to acknowledge the damage and death. Sometimes the animals have the last laugh; the impact of a deer on a small car can be dangerous to the driver as well. Everyone in my town knows the story of the farmer who struck a deer dead-on—elevating the animal just enough to lift it over the hood of the car and sending it through the windshield headfirst—and was impaled in the neck with the sharpest antler point. But the damage didn't stop there. The momentum of the deer carried it well over the front seat of the car through the backseat, and smashed through the rear windshield, meanwhile having completely decapitated the driver! When rescuers reached the crash, the deer could not be found. They assumed the deer was well enough to hobble away. If it eventually died from its wounds, it did as much damage on its way out of this world as it received in the brutal collision.

When I see a dead deer, I don't feel a deep sense of sadness for the poor helpless animal. Usually my sympathy is equally balanced by the possibility of the driver having been decapitated. In other words, the tragedy of the dead animal doesn't overwhelm my emotions. I see the roadkill as evidence of two populations, human and deer, each going about its own selfish ways. Sometimes these populations interact; sometimes the results are violent. Even when individuals are killed in the process, both populations adjust to the interaction. Drivers are more cautious and aware of the deer, and the deer population has one less individual to breed and pass along the dangerous behavior of grazing near highways. It's simply the winnowing effect of natural selection.

I also realize that if the deer population grew out of control, there would be many more accidents. Eventually the humans would have to adjust, by building fences or erecting deer-repellent barriers on the shoulders of the roads. Conversely, if the human population grew, as in large cities, the deer population would shrink drastically. Habitat would change from grazing lands to cemented pavement, and the killing rate would increase from the expansion of cars and trucks on the road. Eventually the amount of roadkill would reduce to almost zero, not due to safer behavior on either side but due to few deer. It's a rarity to see any dead deer on the roads of Los Angeles because the deer population is now sparser than it was before the urban sprawl of the twentieth century. Hardly any drivers in LA have to be mindful of deer as a potential hazard.

The point I am making is that seeing a dead deer on the side of the road isn't so much of an individual tragedy to me as it is an indication about the realities of where I live. I use this information to make sense of my environment instead of getting overwhelmed by emotion of this or that incident. It's emotional detachment in favor of intellectual processing and planning.

I apply this detached attitude to other aspects of my day-to-day experience. I went to college in Los Angeles in the eighties; every day I would pass a few homeless people living on the streets between my off-campus apartment and UCLA. Today, remarkably, homelessness has spread outside our major cities. Now when I tour or travel, I see homeless youngsters begging for cash, and young mothers holding their hands out for change in the suburbs and small towns across the country. It's no longer easy to avoid the realities of a growing homeless population in our land just by moving away from the urban centers.

Still, I try not to get wrapped up in the tragedy of each individual; instead I look at it as a consequence of something affecting our population, and wonder why homelessness became their best—or only—option. The reality is that the absolute number of unemployed people has gone up in our country as the population has increased drastically over my lifetime, and the number of social services has gone down in the past forty years. The cultural environment in the United States has shifted, it seems, and hardened into one that promotes a less compassionate view of our unfortunate citizens. Some assume that homeless people are homeless because they lack the moral backbone or ambition that propels the rest of us forward. Some successful people invoke a kind of pernicious Darwinian misapprehension to justify their negative view of others: If Darwin proved "survival of the fittest," then the less successful people must be less fit and therefore undeserving of successes. The reality of homelessness and indigent citizens is more complicated than a shorthand phrase for evolution, however. Poverty, mental illness, lack of human services, and family or personal tragedies all play a role, and they have more to do with historical circumstances than any personal strengths or failings of the individual members of the homeless population. The political agendas of both dominant parties, particularly in the last decade, have been hostile to government agencies that could help the population of homeless people. Sympathy is out of fashion, as evidenced by cable news, where you are more likely to hear that "we are all on our own" and that government assistance—whether it's funding for homeless shelters, food banks, or mental health facilities—should be ended. Luckily the political environment is changeable, and this can lead to cultural change.

In simple terms our society can be divided into the haves and have-nots. I'm fortunate to be a have, but I still find it easy to sympathize with the have-nots. I know that my future is tied up with theirs, and that making things better for our society as a whole is more important than simply making it better for my immediate family and me.[20] Unfortunately, another view that is common among many of my fellow haves, is a senseless antipathy and narrow-minded harshness toward the less fortunate: These homeless bums are just looking for free handouts, taking advantage of charity, and not carrying their own weight. I had to work hard to get to where I am; no one gave me handouts. This shallow and inaccurate mantra reflects a portion of our society that has ancient roots. It is, like so many unexamined beliefs, a narrative of self-sufficiency that has very little basis in reality. Like many successful Americans, I had an average middle-class upbringing that included an adequate school, a reasonably stable family, and good health. Who knows what would have happened if I'd gone to a shitty school, had unreliable parents, or had a chronic illness. Perhaps I would have become one of the have-nots after all.

The reality is that most of us lucky enough to be haves have enjoyed subtle advantages throughout our lives. If we accept this, and remain aware of the differences between our social situations and those of the less fortunate (and how those helped determine our current status), we can begin to develop a much-needed sense of empathy toward the larger world.[21]

The deer that graze near the highways around my house have no idea that their lives are in danger from the trucks and cars speeding alongside them. Had they been born with an avoidance behavior that kept them away from cars, they could take action to prevent themselves from coming to a potentially violent end. Similarly, without any understanding of the multifarious causes that lead to homelessness, how can a population of privileged but unaware humans develop any kind of public policy or social action to make the situation better?

There's a bigger point here: If we want to avoid extinction, and commit ourselves to long-term sustainability, we will need a greater sense of awareness of what's happening to our environment, on both a local and global scale. We need to learn to care about issues that don't directly affect us as individuals; in other words we will need to worry more about the common good than our own interests. The most obvious way to do this is to reject the superficial and selfish mantras about "makers" and "takers," and instead focus on sharing fact-based knowledge. As social scientists collect more information about the world, for instance, it should become clear that assistance for the less fortunate benefits us all.

For most of the history of Western civilization, life was explained in terms of purpose. Aristotle, the great "first philosopher," began a tradition of teleology that is still the foundation of how we view the world. Teleology is the intellectual practice of explaining things with respect only to their purpose. A teleological explanation for the variety of trees in a forest might go something like this: The purpose of the forest is to provide nesting places for spotted owls and other species of birds. Teleology puts the emphasis on why something exists. In many religions God becomes the answer to most of the "why" questions. Why do I exist? Because God created me for a specific purpose.

Evolution, however, is explicitly nonteleological. There is no ultimate purpose to why you exist, or to all those trees in the forest. The variety was produced over a very long time by "natural laws," including natural selection, heredity, and genetic recombination (among others). Humans, like tree species, are the product of millions of years of evolution and natural selection, and we have extinct ancestors to prove it. With no ultimate-purpose, "how" we came to be becomes a more interesting and relevant question than "why" we came to be.

Part of the problem with teleology is that it biases the average citizen's thinking about social issues. Perhaps you believe that the purpose of a free society is to allow individuals to accumulate as much material wealth as they want, regardless of the cost to the biosphere. If you meet someone who doesn't believe this, then they are in conflict with the purpose of a free society, and they may be ignored, discredited, or worse.

If our purpose is a foregone conclusion and there is an overriding belief that the world exists solely to serve human needs, then we are justified in scorning those who seem to be going against purposeful order. If society is ordered simply to satisfy some overarching purpose then it becomes logical to blame and punish those who challenge the order. But what if our perception of the purpose is wrong? Isn't it just as logical to believe that society, like virtually all populations, is a complex of interacting individuals focused on short-term needs? In such a world there is no single overarching purpose that applies to everyone.

Aristotle believed that the only explanation for animals and plants was in relation to their purpose. "Nature does nothing in vain" is a common quote from his philosophy, in which nature was all-purposeful. All things functioned with respect to its purpose, in deference to a "final cause." But today we know that nature does all kinds of things in vain. Think of mutations, for example. The natural properties of populations reveal no ultimate purpose, but rather they could easily be seen as mere reservoirs of biological and behavioral diversity. We have to conclude, therefore, that Aristotle was wrong in one respect at least. His teleological philosophy cannot easily be applied to humans. If there is a purpose to be had for the human race, it is one that will have to be agreed upon, such as the goal of the longevity and sustainability of our species. This agreement of working together toward an arbitrary goal seems healthy and good for us, but it is entirely different from the "final cause" that Aristotle believed was an inherent property in the universe, responsible for the creation of all things.

Under Aristotle's teleological worldview, all you had to do to figure out why something existed was to figure out its purpose. Once you knew the purpose of something, you then knew the reason for its existence. This idea is a powerful one; throughout Western history things were explained in relation to their purpose. This became the norm in the universities of the Middle Ages, and it was readily adopted by Christian theology as well.

As the Christian population expanded, so did the intellectual idea of ultimate purpose. Most of the activity that took place in universities until the advent of science in the sixteenth and seventeenth centuries was centered on disputation: reconciling discoveries with what was written in the Bible. God's plan, as revealed by his word in scripture, became the go-to explanation for the purpose and existence of all things. Hence Aristotle's "final cause" was repurposed in the universities of this era and understood to be "God's will." Why do species exist in their current forms and seem so perfectly adapted to their habitats? An early naturalist from this period of time would say, Because they are conforming to God's plan.

As far as can be determined from a study of nature today, however, there is no overarching design or plan. Modern scientists are more interested in looking at process rather than purpose. We look at a population and ask how it came to be, instead of the teleological explanation of why it exists.

As soon as you adopt a scientific or naturalistic worldview you realize that populations come together because of properties that are fundamental to their biology. Certain needs must be fulfilled in order for an individual to remain viable. If those needs are not met, existence is impossible. All populations are therefore the products of only "chance and necessity."[22] There is no overarching purpose to their function or existence. This is an explicitly nonteleological way of looking at the world. The contact between any two populations is based on mathematically determined probabilities that depend on highly complex variables, such as the population growth rate, increase in geographic ranges, and resource abundance and efficiency of its utilization, for each population. Once you take into consideration the plethora of groups coexisting simultaneously you realize that these interactions are too complicated to be explained as the result of a simple purposive design. In other words, giving population interactions teleological explanations is overly simplistic and not in accordance with what we see in nature. Instead it is more important to figure out how the two populations came together, and recognize how their histories impact their present state of interaction, with a mind to managing their future.

In general I think it is more exciting to ask how something happened rather than worry about why it happened. Unfortunately most people still think and act teleologically. One of the most popular books in recent years is Rick Warren's *The Purpose Driven Life,* which suggests that everyone is put on Earth by God to fulfill his given purpose. It has sold well over 30 million copies! I find this disappointing but understandable; no doubt it is reassuring to believe that "everything happens for a reason." But the truth is that there is no ultimate reason, and the greatest challenge to overcoming teleology is the admission that our lives have no predetermined purpose, and that, furthermore, we have far less control over the direction of our individual lives than we think. The religious deal with this lack of control by telling themselves that God "has their back." For them it's soothing to believe that God has a purpose for all of us even if we don't know what it is. Believing that God's plan is a secret known only to him might seem comforting to some, but to others—myself included—it brings no sense of peace. I'm a skeptic. I need to see some proof that someone gives a shit about me. I feel that it is far more realistic to assume that no deity (or any corporation for that matter) cares a hoot about me as an individual. I'm content with the love of my family and friends.

I can see how a belief in God and a defined plan makes people feel great. Unfortunately the belief in this "ultimate purpose," and the promise of a paradisiacal afterlife in a supernatural world of endless bounty, stands in stark contrast to the naturalist worldview that sees our biosphere as fragile and finite. The only real afterlife, to a naturalist, is the paradise around us—one that if we are lucky we will be able to leave to our children. There is no real agreement on what is best for the long-term success of our species, and in part this is due to the seductive and reassuring teleological belief system that is at the core of most religions. Each one presents a different path to salvation and a different purpose for the use of Earth for their benefit alone. Many people seem to believe that God's ultimate purpose includes extracting, harvesting, and generally obliterating any life forms, even other humans, that are foolish enough to come between them and potential profits.

The teleological view is closely mated with the persistent belief in free will (discussed in chapter 9). When belief in an ultimate purpose is backed up by a belief that all people are free to pursue it, two very toxic fallouts result: blame and punishment. Why would anyone use their free will to disrupt God's ultimate purpose? The belief is that no one of sound mind should escape being punished if they knowingly do something bad. Leaving aside the data that shows punishment doesn't work,[23] this kind of reasoning is the justification for viewing Rodney King as a perpetrator in the LA riots. Sure, King might not have been behaving like a model citizen when he resisted arrest, but there were profound historical circumstances at play that are far more influential than his supposed free will. Remember that he was driving drunk and resisted arrest, which led to his beating by cops who were probably provoked by his belligerence. The more complex and perhaps relevant "how" questions, however, didn't get asked in that moment. A video captured a portion of the beating, the national news programs played only the most sensational part, showing the swinging batons (not the part where King was belligerent), and the public reacted to the viciousness. When the cops were acquitted of wrongdoing, Los Angeles citizens, mostly from the African-American community, blew off their anger in a fit of violence, vandalism, and terror. The rioters blamed the cops, the cops blamed Rodney King; both groups felt that the other was wrong, the anger and frustration was exacerbated, and the situation quickly devolved into chaos. What should have been at the forefront of everyone's mind was the environment of hostility between the LAPD and the African-American community, which had existed long before the beating took place.[24]

Violence and chaos might be curtailed, sympathy and empathy might become more prevalent, if interacting populations encouraged a wider understanding of one another's history. Context makes it difficult to interpret history simply in terms of "good guys vs. bad guys." But it takes time, sometimes generations or longer, for the context to be understood and spread as general knowledge in a population. Such cultural assimilation, however, might be fostered by an outright rejection of teleology.

Nonhuman species interact only within the context of ecology. Their geographic ranges—the front lines of their population wars—are determined by niches comprised of multiple dimensions (mathematical abstractions referred to as "n-dimensional niche spaces"—more on this later). Most species have no way to avoid the inevitability of conflict if their population is growing out of control. Soon their resources will become scarce and their numbers will dwindle. Once you accept that species in the biosphere have no ultimate purpose—only a proximate purpose under the constraints of environmental selection—you can start to appreciate what brought them together. Scientists often ascribe purpose when they speak of adaptations "for" this or that particular function (for example, flying or swimming). This can be easily misconstrued and equated with theology—that is, how theologians speak of God's plan in the creation myth of the human species. This way of thinking is merely meant to be descriptive, and doesn't square with a starker interpretation of biological data. When taken at face value there is no ultimate purpose for the existence of a species. That is to say there is no long-term, goal-oriented, preferred endpoint, or an intelligently designed trajectory, to any evolutionary lineage or population.

Species exist because of environmental parameters that "allowed" their evolution. Each individual is simply meeting the conditions of existence in order to survive and reproduce, both of which can be considered proximate purposes (those that serve the immediate needs of an organism—nourishment, nesting sites, reproduction, and so on). No biological evidence exists that shows an ultimate purpose to the evolutionary process. In fact evolutionary lineages have repeatedly shown historical trajectories that change direction, speed up, slow down, and sometimes inexplicably terminate, all in response to their ever-changing environment. Studying the fossil record, where data like these are abundant, gives ample reasons to view life as ultimately purposeless.

While this might seem disheartening to some, for me this knowledge gives life more levity. In a simple, everyday sort of way, I can reason that there's no ultimate purpose to a mosquito, so it's futile to hate them for ruining your camping trip. Instead, why not try to understand how mosquitoes function in nature and how they interact with us. It's more productive to think about their general behavior, distribution, and possible breeding sites than it is to complain about the bites. This knowledge allows us to tweak certain aspects of their biology in an effort to actively manage their population and possibly reduce the spread of malaria, the disease for which they are known to be the main vectors, and more prosaically to avoid them while camping. .

Naturalists and environmental scientists look at environmental factors and species' functional ecology first; they don't worry about why species behave the way they do. Instead they simply deduce how things came to be. We need to adopt the same approach when dealing with one another. First and foremost we should come to terms with the fact that we are all of the same species, regardless of nationality or belief system. We form a very large, global population. If we can resist the urge to blame and punish one another, we will improve the odds of our species' long-term survival, and reduce violence in the short run. But to reach this level of enlightenment, we need first to reject teleology, as explained above, and then come to accept some of the basic biological facts about populations.

Most people think of "the population" as implicitly a human construct. For instance, "The population is growing in Jacksonville, Florida" because people are moving there at a rapid rate, making conscious personal choices to relocate: more jobs, better schools, nicer climate, and similar reasons. But biologists use a more basic definition of populations: A population is a collection of organisms linked unconsciously to one another due to some property of their biology—most important, reproduction. The most obvious thing that links all members of a population is heredity—they all carry the same DNA, the same genes. In fact, populations are often defined precisely by the genes that their members all carry.

For instance, the species *Rana pipiens*—the cute little frog that inspired Kermit from *Sesame Street*—is distributed throughout North America, but is divided into numerous discrete populations that function separately. This is common of many widely distributed species: They can be divided up into "races," sometimes called "demes"—groups that differ only superficially from one another. Although all individuals of the species carry the same genome, their genes are variable across the geographic range. In the case of frogs, we can assume that their breeding circle in Oregon is completely isolated from individuals in Florida because it's impossible for a frog to traverse that much distance in its lifetime. Sometimes this is referred to as a localized population.

Localized populations are effectively separated from the "gene pool" of the entire species. A gene pool is a collection of all the gene varieties that comprise a species. Theoretically speaking, all the genes in a gene pool make up the evolutionary raw materials from which natural selection can function. In reality, however, evolution occurs in localized populations, and these are disconnected from other such members of the species. In this sense populations grow and adapt based on their own properties, sometimes on a separate evolutionary trajectory from the other members of the species, because there is effectively no connection to the rest of the gene pool. But again theoretically speaking, if they are the same species they could breed with one another if collected and taken back to the lab.

The various demes that occur throughout a species' range each have slightly different genes. Evolution is defined as a change in the gene frequency from one generation to the next, and this underscores the basic point that most evolutionary biologists like to emphasize: Genetic change—and thus evolution—occurs in isolated populations of a species, and only by adding up all these isolates, and tracking the changes in their gene frequencies from one generation to the next, can you tally the evolution of the species as a whole. Sometimes the isolated populations change in divergent directions, and over time the split creates two distinct new descendant populations. These groups eventually become reproductively isolated, which means that they can no longer interbreed with one another, only among themselves. This is the typical example of one ancestral species splitting into two daughter species.

Some genetic changes get weeded out over many generations because they impair the reproductive potential of the individuals that carry them. Other genetic variations, however, increase the likelihood of successful reproduction, and over time, and through many new descendant generations, they spread to more and more individuals until the frequency of those favorable genes in the population reaches a maximum.

Obviously the frogs are unaware of this process. The individual *Rana pipiens* have no ability to figure out if they are carrying genes that increase or decrease their potential to reproduce successfully. They're simply preoccupied with being frogs. Just as no human instinctively knows if he or she is carrying genes that aid in successful breeding. One thing is certain: It takes a very long time, perhaps thousands of generations, to determine which genetic varieties are on their way out of a population and which are on the way to becoming fixed in a population. Evolution is hard to see in most animals and plants because its time frame is rarely experienced in a single human life. We have no direct experience with the phenomena of long-term evolutionary change. We can only infer its causes, as we witness small slivers of time and small fractions of populations in a continuous process that spans millennia.

Every year I lecture to undergraduates at Cornell University. The class is open to students from every walk of life, which can make teaching evolution very challenging. Most of our students are "nonmajors" who have only a basic grasp of biology or science in general. Every year I get a handful of students who are self-proclaimed intelligent-design creationists. This usually has no bearing on their performance in the class, although they get a bit frustrated when it comes to the lectures and readings about human evolution.

I ask the same thing of all students: Learn the material and think carefully about your reasoning when answering questions on the exam. I'm not trying to convert them. I just hope that they take some time to think about the ways in which their beliefs might be incompatible with science. I never give someone a lower grade because of his belief system. If he rejects evolution as a worldview, he can still pass the course because students are asked only to evaluate the evidence and write about the "organizing principles" or theories that are supported by the empirical facts.

It is a basic fact of science that evolution happens. The simplest definition of evolution is "descent with modification." But that phrase has a profound meaning. Superficially it just looks like "change through time"; and that would not be a bad first impression of evolution. However, the word "descent" means that evolution is dependent on heredity—relationships of parents, offspring, and the long lines of familial connections that stretch through the generations. This long, slow progression of change through time puts the focus on the group as opposed to the individual.

Think about the evolution of the horse, for example. Its journey has been captured in the fossil record. We can trace the horse back roughly 55 million years, and horses show a hereditary link with ancestors that stretch farther back into deep evolutionary time. The oldest horse, *Hyracotherium*, was the size of a small fox. Horses alive today (*Equus*) are more than ten times taller. Height isn't the only anatomical characteristic that has changed. Its teeth are longer and show a more intricate grinding surface than their ancient ancestors'. Modern horses stand and run on one enlarged middle toe on their hind limb and a single middle "finger" on their forelimb. The other toes and fingers are absent, except for vestiges of ancestral structures. We can see where those vestiges originated by studying its ancestor *Hyracotherium. Hyracotherium* had four digits on its forelimb, and three on the hind limb that touched the ground and supported the animal's weight. The key to understanding evolution is understanding how these changes occurred over tens of millions of years; how, in a single lineage of descent, a perfectly formed foot with four toes could change over millions of years into one with only a single digit.

Charles Darwin wrote about this philosophical puzzle from 1838 until his death in 1882. He didn't have the fossil record of horses at his disposal, or any understanding of the hereditary material (DNA was discovered in the twentieth century), or any knowledge of molecular biology. However, his interpretation of nature was so astute that he crafted the modern explanation of how populations diverge from one another and from their ancestors. His keen observations focused on two main aspects of natural history. The first was the recognition that offspring resemble their parents, and the second was that all members of a population show variations in observable traits. To Darwin these two basic facts went hand in hand, and he used these basic observations to infer that all organisms are linked by the same principle of heritable variation. That is to say, he realized that traits are passed along more frequently from parent to offspring, but are imperfect copies of one another. The more favorable traits, those that aid in reproduction and survival, get passed along and proliferate among descendants. The less favorable varieties result in fewer offspring, and lower incidence of the traits in future generations. This process provided Darwin with a deeply significant insight: All organisms are connected by descent, and thus we should be able to trace back in time all organisms to a single original ancestor.

Today the modern theory of Darwinian evolution has been substantiated on so many levels that most educated people accept it as fact. We know that the heritable material is DNA, a large molecule composed of discrete regions known as genes. DNA is found in the nucleus of every cell of your body, and most important it is found in the cells that are stored inside your sex organs. When you reproduce, your offspring inherit a copy of nuclear material from Dad's genes and Mom's genes. The reason offspring don't look exactly like their parents, however, is that they are a mix of materials from both of the parents' sex cells. These are called the "germ" cells. They are found in the gonads—testes in the male, ovaries in the female. They are distinguished from the "somatic," or body, cells. Germ cells participate in heredity, while somatic cells do not. You can change a lot about your body cells throughout life—think about calluses or big muscles—but you cannot change your germ cells. A single egg cell and a single sperm unite at fertilization. These two individual cells carry all the genetic variations necessary to ensure that the offspring will not be an exact clone of either parent.

Variation in the genes assures that the process of sex will always produce offspring that are slightly different from their parents. Another commonly repeated phrase in science is "Mutation is the engine of evolution." Offspring differ from their parents because of changes in their genetic material. Some of those offspring might vary in size, or have slightly different teeth, or some might have more drastic changes, such as the number of fingers or toes.

The parent-offspring hereditary mechanism (which includes genetic crossing-over—or sharing of slightly different genes on chromosomes before they get sorted into sperm or eggs, and low-frequency favorable mutations) ensures variation within the family. It does not ensure, however, that a descendant population will differ significantly from the ancestral population. For that you need some sort of selecting mechanism that will weed out certain varieties of offspring and prevent them from reproducing. This mechanism is called natural selection. Evolution occurs when certain offspring varieties are favored over others across an entire population. We say that those favorable variants have been naturally selected, and since the unfavored fail to pass on their genes, the population that results after another round of breeding will show genetic variation that is different from the previous generation. These naturally selected, slight variations in descendants are the measurable effects of the evolutionary process.

In nature things are much more complicated than this rough sketch of natural selection. The reason that natural selection is so hard to measure in nature is that it depends on two equally important factors that are constantly in flux: (*1*) Heredity—sperm cells are continuously produced throughout life, and each one of them has pluripotential genetic material that is active during millions of cell divisions after reproduction. At each step of embryonic development there is the potential for mutation in one or many of the offspring's cells. Eggs are produced in great quantity, each one of which carries its own set of gene variants. (*2*) Environment—at every moment the environment changes, with each gust of the wind, each encounter with another individual, each flash of sunshine. It is the interaction of these two constantly changing factors that determines the reproductive success of an individual. The cumulative outcome across the entire population of offspring is how we measure natural selection.

This simplification is useful because it demonstrates how natural selection is a two-step process. Individuals can get only so far. Without the appropriate environment as a selecting mechanism, it doesn't matter how fast you can run or how quickly you can solve a problem. Possession of individual traits does not guarantee their long-term success. Reproduction and functional utility are both subject to circumstance. If conditions that promoted success in the past change significantly, a trait that was beneficial in the past may become a burden. This justifies a strong emphasis on understanding the environment's role in shaping the population.

The degree to which an offspring population can vary from its parental stock is predicated on the stability of the environment. If the environment has undergone a drastic change beyond that to which the parent population was finely adapted, the more similar offspring—those resembling their parents—will be selected against. Some fraction of variants among the descendants might match the new environmental parameters well, in which case they will form a new population radically different from their parents. In this case (drastic environmental change in one generation) descendants would not resemble their parents. It is in this sense that we can say that the environment "allows" new varieties to form.

Sometimes the hereditary raw materials, DNA molecules, undergo a biochemical "mistake" called a mutation. These kinds of biochemical mistakes happen within the nucleus of the sex cells when the maternal and paternal genes combine to create a new life. These genetic mutations occur spontaneously and most often cause serious problems in the developing embryo. Although we can think of mutations as a "natural" spontaneous biochemical process, it has been well documented that human activities can increase the rate of mutation through "unnatural" means, such as proximity to radiation, smoking, unhealthy diets, pesticide ingestion, and soon. Even without these human sources, small mutation events occur constantly in the gonads at a theoretically determined rate. Often they go unnoticed because the sex cells fail to become successfully fertilized.

These mistakes are almost always detrimental, causing impairment or death to the individual. But occasionally mutations turn out to be useful in some way. Think of a mutation that caused more hair to grow around the face in a cold climate, for instance. This would be a detriment if the climate turned hot—you would want to shed heat, not retain it. But in a cold climate the extra insulation from a beard could give the individual an edge; he would live longer, reproduce more, and eventually natural selection would bring about a population of furry-faced men. Over the long run of the evolutionary time-scale, a single hereditary mistake in one individual, if beneficial, can affect an entire lineage of organisms because it confers a reproductive advantage to all offspring who carry it.

Horses evolved over 55 million years because of very slight mutations. A male *Hyracotherium* may have passed along a sex cell that had a mutation that caused a slight reduction in all but the middle toe. Somehow this trait benefited his offspring—perhaps allowing them to have a longer, more unobstructed gait. This next generation of horses, being better at running, was able to avoid predators, attract more mates, and breed more successfully than those with a full complement of long toes. The mutation was passed along to every horse descended from this hypothetical ancestral population. Their descendants increased in number until every horse carried that mutated gene. The mutation may have started off in a single individual, but it soon became a fixture in the descendant populations because it was passed along so successfully.

Evolutionary science is essentially the study of populations through time. When we look at our world through an evolutionary lens, we can tease apart hereditary and environmental phenomena and start to figure out the processes that have shaped the human population and those of other species as well. Individuals are not strictly in control of their own destiny, because they are at the mercy of their heredity and their environment. The environment of an organism consists of two parts: (1) Biotic interactions—encounters between organisms; and (2) abiotic factors—those that depend on the climate, geography, altitude, or other physical conditions such as temperature, sunlight, salinity of water, or soil chemistry. The environment plays numerous complicated roles throughout the life of an organism and throughout the history of a population. In humans, culture has a profound influence on biotic interactions. Heredity is the transmission of traits from parent to offspring. The laws of heredity (worked out originally by the Austrian biologist Gregor Mendel) were determined in the absence of environmental influence, under controlled experimental conditions.

In nature the interaction of environment with heredity determines which genes are expressed and in what ratio. The environment also plays a key role in determining probabilities of successful mating. A particular gene might have a very low probability of getting passed on under certain environmental conditions, but if the environment changes the probability of that same gene being expressed might increase. In this way the environment works in tandem with the hereditary process. The flip side of this fact is that individual actions do not shape a population unless the environment allows them to flourish. This is such an important point that it cannot be overstated. If the environment is not receptive to a mutation, it will be quickly eliminated from the population. As a parallel phenomenon, in culture, ideas only spread if they are received by others and then repeated. The cultural environment, therefore, "allows" ideas to spread.

Those of us who live in relative comfort in the Western world like to think that we are in control of our destinies. We watch what we eat, how we exercise, and how we deal with the world around us. Most people assume that evolution is no longer affecting human beings. "After all," they say, "haven't we effectively insulated ourselves from the harshness of natural selection through the buffer of urban living and modern technology?" But evolution is ongoing, and our environment continues to interact with mutations in our genes. Sickle-cell anemia, a genetic affliction that is present in people of African, Middle Eastern, and Mediterranean descent, is an example of this. Sickle-cell anemia causes severe blood problems in those who carry the genetic disease, yet it persists. Why hasn't natural selection weeded it out of the human population entirely? It takes only a few generations to remove deadly mutations under natural circumstances. There must be some selective benefit to having the condition. Let's examine this more closely because it tells us something about how natural selection brings about the coexistence of a perceived evil—a mutation—in our population.

Malaria is a good candidate for being the deadliest disease on the planet. It killed roughly seven hundred thousand people in 2010, and the World Health Organization estimates that one child a minute dies because of the disease in Africa. The prevalence of infection by malaria is closely matched with the prevalence of sickle-cell anemia, in tropical, wet climates where mosquitoes are most abundant. Mosquitoes transmit malaria when they bite, attacking almost all humans living in tropical climates, but especially those at their most helpless: infants.

A microscopic parasite, *Plasmodium falciparum*, is the agent of the disease, and it spends its life cycle in the blood of humans and mosquitoes. After a human is bitten by an infected mosquito, the parasite quickly finds refuge inside the human liver and eventually comes to reside within the circulating red blood cells. If you are carrying this organism around in your bloodstream, you are infected with malaria, and it causes a biochemical change to your red blood cells: They become sticky and clog the arteries that feed your vital organs.

Malaria is deadly. The parasite evolved to travel easily between mosquito and humans; once bitten the victim quickly grows sick, experiencing fever, internal organ dysfunction, and vomiting. Young children die quickly. Human babies, in the time before modern medicine, had no way to resist or combat the infection. It is here that mutation comes back into our discussion. Imagine a mutation that changes the susceptibility of blood cells to infection by the parasite. This is precisely the case in sickle-cell disease.

Individuals with this hereditary mutation are protected from the parasitic malarial infections. It should come as no surprise, then, that the sickle-cell gene is more common in human lineages that originated in tropical climates where malaria is most common.

The sickle-cell trait stems from human mutation on one portion of a single gene, and it's relatively benign if it occurs on only one of the paired chromosomes in an offspring (that is to say if the child received the gene from only one parent). The offspring in this case are considered "carriers" of the sickle-cell trait. If, however, an offspring inherits two of these mutations, one from Mom and one from Dad, very serious problems arise.

Sickle-cell disease shows up as improperly formed red blood cells. These cells are responsible for carrying oxygen to all the tissues of the body, and in sickle-cell disease they become flattened and curved like a sickle. This makes them somewhat immune to the malarial parasite. But there is a downside. The malformation also causes obstruction of the blood vessels and problems with kidney function, lung function, spleen function, and a host of other serious problems. Carriers of the disease might live long enough to reproduce and live fairly normal lives with minor ailments coming from the malformed red blood cells. People who inherit two copies of the sickle-cell gene variant (one from each parent), however, will experience more severe cases of the condition and die earlier having had fewer—if any—children.

The population of vulnerable people in the tropics is thus stuck between two diseases, one caused by heredity (sickle-cell), and one caused by the environment (malaria). This balance between two terrible endpoints—sickle-cell anemia or malarial parasite—is called an evolutionary trade-off. An evolutionary trade-off is a compromise between two lethal alternatives. Both alternatives will lead to the individual's death, but one (sickle-cell trait) has a built-in mechanism that will allow the individual to live long enough to reproduce.[25]

People who carry the sickle-cell gene variant might pass it on to their offspring. As long as the mutation allows the offspring to reach reproductive age, its presence will outstrip the negative effects of malaria by eliminating the "normal" red-blood-cell host preferred by the parasite. The long-term result will be growth of the human population in the presence of malaria because more offspring are living to reproductive age. Eventually this will result in a balanced frequency of the sickle-cell mutation in the population. Hence there is "just enough" of the mutation present in our species to match the proportion of those living in the presence of malaria.

The point of all this is to illustrate that everything in life is influenced by both hereditary and environmental factors. From such a perspective, we can perceive that much of life is neither completely within our control nor entirely our fault. And it serves to illustrate the difficulty of pinning the blame for a problem on a single culprit. For instance, if we destroyed all the carriers of the sickle-cell trait the environmental problem would still remain: mosquitoes carrying *Plasmodium falciparum* in the tropics. Infection rates would soar among those humans who live alongside malaria-causing mosquitoes because we would have removed from the population all those who carried natural resistance to malaria (those with the sickle-cell trait mutation).

Mutations arise spontaneously; the only way to prevent the spread of mutations is to limit reproduction. This idea comes worryingly close to eugenics, but—whether we want to admit it or not—a form of eugenics is happening all the time, all around us. For example, most parents are now offered the chance to screen their unborn child for Down syndrome and other prenatal maladies. Some parents, upon learning of a possible mutation early on in the life of the embryo, choose to terminate the pregnancy. We can prevent some unwanted mutations from spreading, but we can't prevent mutation itself because it is a fundamental biochemical process that goes on during the development of every sperm and egg.

We can apply this outlook to our cultural environment. The political process, for instance, is rife with failed experiments and shortsighted schemes. I like to think that great ideas can come out of "mistakes" in the same way that genetic mutations are a kind of mistake in the nucleus of a cell. We have hundreds of seemingly useless thoughts every day, which arise out of nothing and seem to have no purpose. Most of these "mistakes" are forgotten before you're even able to register them fully as ideas. However, some of these mutant thoughts are the spark of something brilliant and big.

Ideas live and die in the same way, and for the same reason, as genetic mutations. Valuable knowledge gets passed on to younger generations, friends, peers, and so on, who in turn pass them along to their kids. Human knowledge is as vulnerable to the forces of selection as are traits in nonhuman species. If the idea is useful (note: *not* the same thing as "good") to the individual it will spread in the population, but a weak or unuseful idea won't last long. Our thoughts are spontaneous, quick moving, and unceasing, so there's no end to the potentially cockamamie mutational ideas we can come up with. My local newspaper is packed with evidence of this. Sometimes these ideas have surprising longevity. Other times they just fall flat.

I consider the idea that "corporations are people" to be a remarkable mutant idea. The idea of corporate personhood has a precedent; the U.S. Supreme court first invoked it when "corporate personhood" had to be established in lawsuits. Corporations are entitled to certain protections if they're being sued, just as a person is entitled to certain protections if she is being sued. But the right-wing group Citizens United, which bills itself as political documentary filmmakers (or, more accurately, political attack ads that are falsely robed as documentary films) decided to try and push this idea further in the Supreme Court. This wasn't supposed to happen. In 2002 a bipartisan law banned attack ads during the election season: The bill enjoyed bipartisan support in Congress. Most Americans from both major political parties agreed that this was a sensible thing to do.

Citizens United protested this law, lobbied heavily to change it, and the right-leaning Supreme Court reversed it in 2010. Their ruling invoked the U.S. Constitution's First Amendment right to freedom of speech. Essentially they said that companies like Citizens United—which receive tens of millions of dollars in corporate donations—were entitled to freedom of speech just like any American citizen. Radio and television advertisements were now open forums for corporations to run anticandidate propaganda (previously illegal under the bipartisan law) just like a street corner where Joe Blow could shout his opinions without fear of legal repercussion. Somehow, Citizens United hoodwinked the Court into believing that corporations should be as unrestricted in their social reach as ordinary citizens; and the Supreme Court created an environment for this mutant idea to flourish. Now the idea that "corporations deserve free speech just like citizens do" is justified by official sanction.

The problem of course, and the grave error made by the Supreme Court (as pointed out in the dissenting opinions by minority judges), was that corporations were not the intended benefactors of the freedoms laid out by the Founding Fathers. The idea that the people should be allowed freedom to express their opinions was not intended for corporations:

". . . corporations have no consciences, no beliefs, no feelings, no thoughts, no desires. Corporations help structure and facilitate the activities of human beings, to be sure, and their 'personhood' often serves as a useful legal fiction. But they are not themselves members of "We the People" by whom and for whom our Constitution was established."[26]

Let's look at this mutated idea from an evolutionary perspective. It originated as a mutation of a preexisting functional idea: American citizens deserve free speech, and therefore you can't put someone in jail for expressing their opinion or for publicly presenting factual information. As this idea spread some people lost sight of what the First Amendment actually protects. Some people sincerely believe that corporations deserve First Amendment rights, likely because of the clever propaganda produced by Citizens United that misrepresented the ideas behind the Court case.

In reality, of course, a corporation wields far more power than any individual. So much power, in fact, that we often require limits on that power to prevent them from abusing it. Corporations don't need the First Amendment; ordinary citizens do. But as the saying "Corporations are people" gets repeated, the environment becomes more receptive and the mutation spreads.

I'd suggest that this process is akin to the concept of horizontal gene transfer. Usually genes are passed from parent to offspring. In other words, organisms have to reproduce in order to spread their genetic material. Most bacteria, however, have the ability to share their genes with others of their own generation. This means that a bacterium's gene for protein A can be given to another mature individual in its population, and that bacterium will then also begin to make protein A. Imagine if humans could do this. We could, for instance, simply change our eye color by receiving a gene from a particularly striking model whose eyes we adore through donation of her eye-color gene directly from her. Of course we know that we can't acquire genes from those we adore. In order to share our genes we need to create a whole new person through the process of reproduction (hopefully with someone whose eyes we adore).

The process of sharing genes between individuals is considered "horizontal" because it doesn't require reproduction "vertically" down lines of descent from one generation to the next. Ideas similarly can be considered as spreading horizontally. They can pass vertically, as in early education or perhaps family traditions, but since we are exposed to new ideas throughout life—and sometimes we incorporate those ideas in our adult behaviors—there is a strong analogy with horizontal transmission. Ideas can simply spread unchecked, from one person to the next, among the standing population. It is this property of ideas that led the English biologist Richard Dawkins to propose the concept of memes. Memes are like genes: They contain information, and they self-replicate. But in order to do so they need human brains and culture in which to spread. Writing, language, and symbolism are the modes of replication of ideas, good or bad. Sometimes they are forgotten—even good ideas—in which case replication of the meme ends. Sometimes bad or pernicious memes spread, not because they are correct: they are passed on simply because they are popular.

But here you must recognize that we cannot do anything about controlling mutated ideas (or memes) at the source because this would require silencing dissent, censoring expression, eliminating free thought, and criminalizing intellectualism—strong-arm practices that have definitively failed throughout history. Once an individual unleashes a novel idea (a cultural mutation) the social environment decides if it is viable. In other words, if we want to do anything about treating a mutation, we need to do the same thing we are doing to treat malaria: tweak the environment so that the idea no longer holds any importance.

Human intervention at the genetic level does almost nothing to prevent the ultimate cause of sickle-cell disease. The only useful intervention is testing parents to see if they carry the appropriate genetic variant. If they both test positive they can take precautions to ensure that the fertilized embryo is healthy. There is a likelihood that sickle-cell disease will be avoided due to mathematical properties of gene assortment in the sperm and egg. Sickle-cell disease is a simple Mendelian trait, which means two carriers of the gene variant will produce only one in four offspring that show the extreme form of the disease. In vitro fertilization of the egg allows doctors to detect the healthy embryos from those that are positive for sickle-cell anemia gene variant. The parents can then decide to implant the healthy embryo into the mother's womb. But this is an individual choice, not a population fix. The parasitic infections will continue to plague the larger community. If we want to make the biggest impact treating any of these diseases we have to intervene in the environment. Managing the breeding habitats of mosquitoes, or reducing the contact between humans and the pathogen (i.e. with mosquito netting), is ultimately more effective than treating individuals one at a time. It may not eliminate the parasite, but it will prevent malaria transmission, which means fewer human infections.

Sometimes, when dealing with mutant ideas, we can adopt a similar strategy. Citizens United is able to spread its ideas because our cultural and political environment allows for it. The Supreme Court gave the members a habitat into which they could release the disease. Until the Court gets back on track and overturns its overturn, we might be in for a sustained period of tragic consequences in our political system. The best way to quell a bad idea such as "Corporations are people" is to expose its proponents and their hidden agenda. Bad ideas rarely spread when the population is educated about better alternatives.

Mutations have a hard time coping with change. The vast majority of them result in the death or inviability of an embryo. If a mutation is advantageous in any way, it will spread through the generations, eventually replacing the original trait from which it was derived. When all members of a population are endowed with the trait in mutant form, it is considered a "fixed" mutation. Once they reach fixation, they depend on constancy of the environment. We don't know the future of challenges to the First Amendment. All it takes, however, is a "viral meme"—perhaps a catchy song, persuasive movie, political speech, or news story—to remind people of the ruthless human toll caused by certain careless corporations in our midst, and the environment may turn cold to the idea that "Corporations are people."

4

THE CONTEXT OF PERSISTENCE, THE BACTERIAL DIMENSION

I wrote a song in 2013 called "Past Is Dead" that paints a picture of my beliefs about population wars. The title of the song is ironic, because the past is very much alive, yet we are often willfully blind to the fact that our past has a huge influence on our present. The first line of the song is "Strewn about the battlefields of life are the remainders of history," and when I look at the world I see ancient history that is very much alive in the present. This is true on both a cultural level and, as we will see, on a genetic and cellular level, within our own bodies. More important, I recognize that the past lives on in a mixture of both the winners and losers of history. We are a global population that is constantly assimilating, and there is no absolute delineation between the populations of the conquered and the conquerors.

I believe that modern humans are poised to acknowledge the cold truth that ultimately there are no victors and no vanquished in life, and that no one is ultimately entirely good or inherently evil. If we, as the intellectual community of modern world citizens, can make this idea accepted, then humans will coexist more peaceably, and we can focus on changing how we deal with other species in the interest of creating a sustainable future.

When I read the history of warfare, I am less interested in the myth of who is the perceived victor than I am in the resulting longer-term assimilations. Yet our culture is obsessed with the belief that winning means complete domination of your opponent, as if war is some kind of competition with a clear prize for the winner. If this is correct, then why do we have winners and losers existing side by side today? In short, it is because of the persistence of populations and the futility of extermination.

In this and the next few chapters we're going to examine this persistence on every level from micro to macro, first looking at how replicating populations such as microscopic organisms (microbes) and viruses work within our bodies, how they have persisted and replicated since the origin of the biosphere, long before our species evolved. We like to think of ourselves as autonomous, self-regulating beings. The reality is that each one of us is controlled in part by other microscopic beings.

For starters, the genetic material we carry is possibly as much as 43 percent viral. In other words, viruses that infected our ancestors millions of years ago have effectively been incorporated into our genome. Clearly some sort of compromise was reached over the years, with the host species providing a haven for viral DNA.

Furthermore, the bacteria in our guts were once thought to be important only in regard to the digestion and absorption of food; but it is now clear that our bacterial microbiota is involved in developing our immune system and is critical to our overall well-being.

In another area researchers are currently exploring how parasites, such as the feline parasite *Toxoplasma gondii,* can actively change our brains and our behavior. The parasite changes its host's behavior to be attracted to, rather than frightened by, cats. Rats infected by *T. gondii* exhibit signs of curiosity, excitement, and even sexual attraction to the smell of cat urine. The virus benefits from the change because it needs to return to a feline host to reproduce, but obviously this scenario does not end well for the rats. Humans infected with *T. gondii* exhibit equally strange changes in personality and behavior (infected men show more oppositional defiance and jealousy; infected women become more warm hearted and moralistic), though thankfully the infection in humans rarely ends with the host being eaten by cats.

The fact that viruses can rewrite our genetic code, bacteria can affect our mood, and parasites can change our decision-making processes and lessen our inhibitions has interesting implications. One possibility is that we might begin to consider "ourselves" as a coalition of many populations, rather than one definitive "I." Perhaps one driver of evolutionary success is how well we maintain our internal environment—how well we care for the billions of microbes that dwell within and on us. We are all stewards of our own unique internal environment; perhaps learning to care for it will encourage us to care for our external environment as well.

The general public easily confuses bacteria, parasites, and viruses (just ask any doctor who has been asked for antibiotics to treat the flu), yet they have little in common beyond their microscopic size and the fact that they impact human lives on a daily basis. Let's take a closer look first at the way bacteria coexist both with us and within us, affecting our external environment, health, and emotional well-being.

My house is built on rural land in the Finger Lakes region of upstate New York. The geology below the concrete foundation is fascinating; a few feet under our house is a thick deposit of unconsolidated glacial sediments sitting atop a huge stack of ancient strata. These gray and black sedimentary rocks have numerous features that indicate an ancient biological signature in the history of their formation. Furthermore, the stratigraphy (the orientation and sequence of the sedimentary rocks) in this region belies a unique geology that reveals the historical setting of our land over the last 400 million years.

Reading history in rocks is like interpreting a map. It requires a good imagination and keen observation. The stratigraphic rock record isn't a map of places, however, but of time. Stratigraphers piece together bits and pieces of evidence found in the sediments to reconstruct the geological events of the past. This information allows them to slowly build a picture of the succession of evolving environments that were laid down, layer upon layer, through geologic time.

Stratigraphers have been studying sedimentary rocks for more than two hundred years. One of their most meaningful discoveries is conclusive proof of bacteria in 3.5-billion-year-old sediments. This shows us that bacteria were the first living things to flourish on our planet. Although bacterial cells are very difficult to find as fossils, the chemical evidence they left behind gives us lasting clues that they were as productive in the past as they are now. Bacteria were everywhere, long before the advent of "higher" organisms like plants or animals.

We have a hay field and a large vegetable garden on our property; we spend a lot of time maintaining and encouraging the alfalfa, tomatoes, and kale to grow. This means that we know our dirt, and the rocks that lie beneath it, pretty well. The sedimentary rocks that form the bedrock under my house, and the surrounding outcrops in the nearby gorges and glens, are gray. The color is important; it gives us an easy clue to the history of the rocks. Black and gray sedimentary rocks were deposited under oxygen-free conditions. No oxygen means no oxidation; i.e., no rust. Other sedimentary rocks (such as those in the Four Corners region out west) are red; they contain hematite, an iron-rich mineral that rusts when it is exposed to oxygen. These red beds react like this because they were formed in conditions of oxygen abundance, in the presence of photosynthetic organisms. Yet no matter what their color, all sedimentary rocks contain the same basic set of minerals. They are, after all, the eroded bits and pieces of mountains, floodplains, or reefs. The big difference is where these "bits" ended up and what was—or wasn't—waiting for them when they got there.

Let's say our "bits" ended up in a shallow marine basin. In today's climatic conditions such places are warm, bathed in abundant sunlight, and are inviting habitats for photosynthetic microbes. These microbes in turn produce oxygen, which can rust iron grains as sediment accumulates. Terrestrial basins, such as those in the deserts of the southwestern United States, also produce red beds because the abundant oxygen in the atmosphere rusts residual iron in the sediments. This can also be seen in tropical soils of the Amazon basin today, as well as in regions of the Sonoran Desert, where seasonal rivers produce a rich reddish landscape.

Deep-ocean basins, however, are a different story altogether. Sediments pour into them from large river systems that drain continental highlands. Deep basins are devoid of sunlight and therefore less hospitable to photosynthetic microbes; this means no oxygen—as we see today in the Black Sea. Sediments pour in, and the finer-grained materials that are destined to become future shales and mudstones remain black and gray as they come to rest in the deepest portion of the basin. Without oxygen no rust forms.

There's another crucial component that makes these deep-water sediments different from red beds: biological debris. Dead photosynthetic microbes accumulate over time and pile up in thick layers underneath tons of sediments in deep basins in total darkness beyond the ocean depth that sunlight can penetrate. The photic zone is only the upper one hundred meters of water depth in the ocean. The sunlight that penetrates, and the warmth of this zone, creates a habitat that is teeming with algae and other small life forms. When these tiny creatures die they drift slowly to the seafloor thousands of meters below. Plant debris from the land and nearshore environment also gets washed into the ocean basin and begins to settle on the seabed. All this biological detritus contains appreciable amounts of sulfur, particularly red algae.[27] Anaerobic bacteria that live in deep-ocean basins use sulfur to create energy. The color of black shales comes from pyrite, a sulfur-rich mineral that forms when bacteria are actively producing hydrogen sulfide. However, this process, like most processes that generate energy, creates a byproduct, in this case hydrogen sulfide gas. Hydrogen sulfide is highly destructive to shells and bones. It dissolves calcium carbonate, so it is not uncommon to find black shales devoid of fossilized skeletal parts. Impressions in the mud might be present, but no skeletons are found in these deep-ocean basins where bacteria are busy churning out hydrogen sulfide.

Black shales are best known to petroleum geologists for one simple reason: They contain fossil fuels. The anaerobic bacteria—aka extremophiles—that produce the hydrogen sulfide consume some portion of the dead organisms for energy, particularly the sulfur. The sulfur comes from the cell walls of red algae, so the bacteria can gorge themselves on that while leaving the lipid portion of their meal uneaten. All cell membranes are made of lipids, and it's this leftover oily portion that collects and becomes hydrocarbons known as kerogen, which infuses the sedimentary rocks. Most oil or gas reserves are a direct product of the molecular remnants of photosynthetic algae and bacteria that died and settled on the seafloor[28] and then formed kerogen. The kerogen, safe on its inhospitable basin floor, remains undigested and unused by other organisms and is eventually converted into oil and gas.

The plants and animals that were compressed into the sedimentary rocks below my house died in the Middle Devonian, 400 million years ago. However I, like many other landowners in the northeastern United States, am still dealing with the direct consequences of those organisms. The lack of skeletal fossils, the black color of the rocks, the horizontal stratification of the glens and cliffs, the smell of hydrocarbon in the well water. All these things are constant reminders of the past worlds that existed in this part of upstate New York, and of the factors I had to confront when our house was being built.

Our recently built farmhouse is the picture of modern convenience, utility, and wise environmental sustainability. Although it is rather large by urban standards, it uses far less energy than most houses half its size. This is accomplished by using the most modern building materials available and heating strategies that weren't very common even twenty years ago. Hopefully they will become standard or be surpassed in the next twenty years.

One of the key elements of our sustainability effort is a fully self-contained water budget. We have a well that draws from an aquifer 120 feet below the surface. Our used water (politely called "gray water") is essentially recycled because it empties from the main sewage pipe into a tank full of digesting bacteria that reside within a thousand-gallon holding tank. When the bacteria have digested all the waste from our household, the gray water spills out into a labyrinth of pipes that extend hundreds of feet away from the house. In this "leach field" the household effluent returns to the soil, where it percolates and the water eventually rejoins the aquifer hundreds of feet below. We also capture water off our barn during rainstorms and store it underground in thousand-gallon containers. All the irrigation of our gardens and lawns and water for car washing comes from our rainwater capture systems.

One caveat about our water, however, comes from the nature of the sedimentary rocks below our property. We live roughly twenty-five hundred feet above the Marcellus Shale, a large Devonian formation that has been in the news recently because of the biological treasure enclosed within its layers.

The Marcellus Shale is one of many sedimentary rock units that stack up like a layer cake throughout upstate New York, Pennsylvania, West Virginia, and Ohio. These rocks were deposited around 400 million years ago. The subsurface beneath our house is composed of neat layers that contain some fossils and lots of hydrocarbons. It has not been disturbed by tectonic forces. Whatever was buried down there 400 million years ago remains.

This is not always the case for ancient sediments. Tectonic movements can bring ancient sedimentary rocks up from the deep and leave them on display for us to view. Some of the best examples of this are seen along Southern California's highways and country roads. I've spent a lot of time driving either to or from Los Angeles. The huge broken slabs of sedimentary rock that jut up along the sides of the mountain highways impress me as much today as they did when I was a teenager, seeing them for the first time. These rocks are still moving; the San Andreas Fault is compressing them at the rate of roughly six centimeters per year, though most Californians choose to underplay its explosive potential as they build their homes and work within this powerful fault zone.

Sedimentary rocks, and their record of the ancient world, are all around us. They can be seen relatively undisturbed in areas that don't experience violent earthquakes or any kind of active tectonic disruption. The Colorado Plateau is a good example. The plateau is like a huge layer cake of rocks three thousand feet thick, which have gently risen without folding, warping, or compressing. Within the last 3.5 million years (recently, by geologic standards) the Colorado River's massive discharge eroded its banks and began to cut through the layers of this undisturbed sediment pile. The result is the Grand Canyon, one of the most magnificent sights in the natural world.

From the rim of the canyon, you can peer down and see the Colorado River three thousand feet below. It's still carving its way through the rock—although now it has reached "basement" rocks so old and changed from heat and pressure that they don't even look like sediments anymore. From your vantage point on the rim, however, you will notice that you are standing on a white-colored sedimentary unit that is also visible miles away across the gaping divide of the canyon. This rimrock, the Kaibab limestone, is like the topmost piece of plywood on a stack at the hardware store. The layers below it can be thought of as sheets laid down in progressive sequence. The absence of tectonic folding and squeezing allowed for the great revealing of undisturbed layers in canyons that were eroded by the river below.

The Great Lakes region is another undisturbed showcase of Paleozoic sediments. I've driven through Wisconsin to central Illinois, across to Pittsburgh and Cleveland, and over all of upstate western New York many times. There is a quieter, homey, more rural vibe here than in the Grand Canyon region. But the rocks are similar in their horizontality.

The forces that shape the Western landscape are big and dramatic; there is something attention getting about the tectonics and mountain building and canyon carving that dominate the vast open vistas of the American West. Back east our landscape is softer and less extreme, in part because our most noticeable landscape features were formed by glaciers. Until twenty-two thousand years ago—during the Pleistocene Epoch—much of this region was covered by a sheet of ice that was more than two thousand feet thick. The winter season lasted longer, the summers weren't sufficiently long or hot to melt all the snowfall, and gradually year after year ice became layered and piled up on the horizontally undisturbed sedimentary rock layers below.

The thick ice sheet wasn't stagnant. Ice moves like water, just much more slowly. It carves its way down a gravitational gradient—from highland to lowland—just like a stream. Glaciers can carve valleys, however, much wider and more rounded in profile than water can. The continental glaciers of the eastern United States carved wide valleys and deep holes for hundreds of miles as they advanced from colder Canadian regions to their terminus in southern Illinois and New Jersey.

As the climate began to warm (twenty-two thousand years ago), the rate of melting exceeded the snowfall accumulation, and the ice sheet began to retreat northward, up to the cold Arctic latitudes where they originated. As they retreated, they left huge piles of debris along their edges. These debris piles are everywhere, forming the rolling hills and skyline ridges throughout the southern Great Lakes region and Finger Lakes of New York State.

Our Eastern geology may be less dynamic than its Western equivalent—we have cozy glens and eroded hollows rather than lofty mountains and fault-bound basins—but it is equally beautiful. Some of the most dramatic scenery in the East can be found alongside the rushing waterfalls and streams that flow through the glacial amphitheaters only miles from my front door. The cliffs along the Great Lakes, and the gorges that feed the Finger Lakes, for example, are lined with sedimentary rock cliffs hundreds of feet tall. A rim of unconsolidated Pleistocene sediment caps these cliffs. Forest trees cling tenaciously to the deep soil that rims the gorges and cliffs. Our farm's immediate area is full of dramatic gorges and waterfalls that plunge down the cliff faces; just down the road, in the village of Montour Falls, the She-Qua-Ga Falls seem to pour over a handful of Early American houses still standing at the base of the cliff. In the winter the falls freeze over, but you can still hear the water rushing beneath them. I have to remind visitors not to get too close to the iced-over creeks and rivers, lest they break through the fragile crust and get swept away.

Two hundred and fifty feet away from our house sits a glen with a raging creek at its base. We are comfortably situated on a rolling meadow; under the topsoil lies about eighty feet of Pleistocene gravel left behind by the retreating glacier that passed through here roughly ten thousand years ago. Drilling our water well was an education on just how much sediment was left behind.

As the drillers plunged deeper into the earth, their steel drill, boring down inch by inch, sent tiny bits of broken rock and debris in its wake, back up to the surface. Solid rock was struck about 78 feet down. Still no water of appreciable volume. They switched to a different drill bit, one that could cut through the solid rock with ease. Mudstone, siltstone, sandstone, more mudstone. Layer after layer they plunged, until eventually, at 120 feet, they struck a high-volume water "seam" and our well was established.

The water slowly rose to fill the hole bored by the shaft of the drill. The drillers sank a steel pipe vertically down the shaft to contain the water as it gurgled up toward the surface. As it spilled out onto the ground, it brought with it fine sediment, clay, and the gaseous remnants of bacteria; a faint smell of methane emanated from the pipe.

Geologic maps of New York reveal that the sedimentary rocks into which our well was drilled continue down in layer-cake fashion. If the drill had bored two thousand feet, it would have reached the "mother lode" of local hydrocarbon lore, the Marcellus Shale.

What makes the Marcellus unique in our region is the kerogen it contains. Kerogen, as already mentioned, is formed from decayed animal and plant matter that eventually accumulated in deep-ocean basins, which, in turn, eventually formed black shale. Think of this shale deposit as a huge compost pile 391 million years old. Conditions were just right to cause a chemical alteration in that decaying heap of algae and plant debris; an alteration that some unimagined future organisms—with big brains, strong backs, and weak foresight—would exploit for their own evolution. Anoxia (the lack of oxygen) prevented the organic debris from breaking down.

Under "normal" conditions, like here on Earth's surface, or in nearshore environments, oxygen is plentiful. Oxygen breaks down dead organic tissues either by allowing aerobic bacteria to consume them, or by enabling an oxidizing chemical reaction. Both of these processes break down organic debris (through oxidation). When an animal dies, the chemical "attack" of oxygen degrades the carcass. Aerobic microbes and other animals can help this process. They eat the carbon and use it to make organic "building blocks" for their biological needs. Oxygen gets released as CO_2 gas in this process of oxidative respiration. So, on the surface of Earth, carbon is constantly being converted in the presence of oxygen, which results in the disintegration of all dead organisms.

A lack of oxygen is a problem for human health, but historical anaerobic conditions have proved to be a boon for those humans who make their living in energy extraction. The Marcellus Shale was the perfect environment for the creation of hydrocarbons. The organic material was buried rapidly, there was no oxygen, and the sediment pile grew thick. With depth comes heat. At just the right depth, and with the right amount of heat, the organic remains might get "cooked" into crude oil. Under other conditions of heat and pressure, however, the kerogen is converted into methane gas. The gas in the Marcellus Shale is still down there today, as it has been for nearly 400 million years, now trapped under thousands of feet of lithified sediments.

Time is important in this process; the sediments quickly accumulated to form the Marcellus Shale. For five million years, dead organisms drifted into the Marcellus basin and were rapidly buried by more sediment. As the basin continually filled up with sediments and dead microbes, the conditions at the bottom changed and were no longer sufficient to "cook" the organic material into oil or gas. Gas production slowly came to a halt. Eventually the sediment pile reached the shallower well-oxygenated ocean-surface waters. By the end of the Middle Devonian (383 mya), the basin was shallow enough to support a well-developed coral reef, which we recognize today as a geological formation called the Tully Limestone.

The Tully Limestone was a fully formed coral-reef community that basked in the healthy glow of the Devonian intertidal photic zone. If we could travel back in time to experience it in its living glory, we might vaguely recognize some of the animals living on the reef. At the very least they would be recognizable as "seashells by the seashore." At the bottom of this pile of Devonian sedimentary rocks, however, is a completely alien world. This environment, like the depths of the Black Sea today, was devoid of any web of complex "higher" organisms, only bacteria that can withstand the toxic conditions.

Anoxic oceanic basins are hotbeds of bacterial anaerobic communities. These populations live in the presence of methane the same way we are surrounded by oxygen. Some give off methane as a by-product, and some eat methane for their energy needs. These organisms were there during the formation of the Marcellus sediments, and they are still with us today, as I discovered the day we turned on the tap water in our new house.

When we first drew up our plans for the house we decided that a reliable well was crucial. However, we got more than we bargained for. We quickly learned that in this region wells often have problems with methane. Also known as "swamp gas," it can smell offensive when you go to wash your face or draw some water for a kettle.

This embarrassing smell isn't the water but instead the methane, the most common waste product of anaerobic bacteria. It's not the bacteria's fault; they've been doing fine for hundreds of millions of years until we came along and invited them into our house through a brand-new well pipe and plumbing system. We can't blame them for expanding their population into this new, wide-open habitat that we created.

The smooth interior surface of the pipes might be glassy and shimmering to our naked eye, but to a bacterium those pipe walls have deep canyons, crenulations, and caverns. Bacteria can easily get into these and form colonies. In the language of microbiology, these are called biofilms, microscopic layers of dead and living bacterial colonies built one generation upon another.[29] Folded and crenulated, these films create surfaces of attachment for all kinds of other bacteria. When water is stagnant, biofilms may become anoxic from lack of oxygen circulation. But when turbulent water disrupts the tranquillity of the anaerobic colonies, oxygen is brought in and the flow washes them away. Some biofilms persist even through relatively constant water flow. But in general, the more water flowing through the pipes, the lower the incidence of biofilm buildup. Still, there are thousands upon thousands of potential habitats that are just perfectly suited for anaerobic bacteria within the pipes leading to your faucets. These microhabitats are where the anaerobic bacteria that release methane reside.[30]

We have one major advantage in this population war between us and the odiferous bacteria—and it's a great little twist, straight out of *War of the Worlds*. The element that we depend on for our most basic environmental need, oxygen, kills our opponents the second they are exposed to it. Anaerobic bacteria die when they come into contact with oxygen. In our house this happens when they are sieved through the aeration screen on the faucet's mouth. So we are left with no threat from their carcasses, simply the unpleasant gaseous emissions from their colony.

Still, they don't go without putting up a fight. Interior plumbing has been a fixture of American life for at least 150 years, so a lot of houses have long histories of infection by methane producers. Some pipes are probably so infused with bacterial colonies that they will never be treatable with conventional whole-house filters. This is why faucet filters that are placed under the sink are so popular—they are farthest downstream, so it's less likely that there will be a big colony in the short run of pipe to spigot, where deadly oxygen reigns.

In our house, however, there is a peculiar phenomenon where some pipes have momentary blasts of methane acridity and some seem pure. This is a puzzle that required some review of our building plans. One day our plumbing contractor finished connecting the elaborate labyrinth of pipes that led to our three bathrooms, kitchen, and laundry room. It was time, he said, to hook up the "main line" water pipe that came from our well. At this initial stage only one bathroom and the laundry room had finished plumbing, and the house wasn't yet protected by a whole-house filtration system. Still, the plumber wanted to test the well pump and get some water into the house.

The water that finally flowed through the first tap was cloudy with clay, and there was a faint smell of methane gas from each of the fixtures. Where did this smell come from? Basically there were two possibilities: The first could be ancient gas from a Devonian population of methane producers. We might have hit a "seam" of gas that had migrated through cracks upward from the Marcellus Shale into our well. Or it could be a biogenic source of methane being produced today by living bacteria much closer to the surface. These populations thrive wherever there are appropriate habitats with low oxygen levels and easy access to carbon (such as the dead piles of leaves, animals, and microbes in swamps or bogs). Either way bacteria, whether living or long dead, were the problem.

So the real question for mitigating the plumbing problems in your house is: Are the faucets emitting a methane smell because of biogenic processes or is it a consequence of tapping into a fossil hydrocarbon reservoir? If it's the latter, you might have to drill a new water well, but you could also get rich if you sell off the hydrocarbon resource to a gas company.

If, however, your problem is from biogenic methane, then you have to control the bacterial population in order to solve the problem. Eventually we bought and installed a whole-house filter and placed it upstream from the pipes in the house. By the time we connected the rest of the house plumbing we could be sure that no bacteria could come into our house because of the physical barrier provided by the whole-house filter. Not even bacteria can pass through a filter with a pore-space diameter of one micron—bacteria are about ten microns wide.

We are lucky in that our infection was limited to two taps in our house. We have neighbors whose hundred-year-old pipes are completely infected. It's unlikely they will ever be free of the bacteria and their accompanying biogenic methane.

Our house-building episode is one tiny incident in an evolutionarily brief moment in time. Bacteria and their close relatives were the first living creatures on our planet, yet we are only just starting to truly understand how they have adapted to our evolution and how they continue to shape human lives. Bacteria were discovered only in 1687, by a Dutch amateur scientist called Antoni van Leeuwenhoek; his unique skill at grinding lenses allowed him to build very early microscopes. He studied samples of tooth plaque and observed "very little living animalcules, very prettily a-moving."[31] Yet it took many years for scientists to begin to understand what bacteria were and how they operated. It wasn't until 1850 that a Hungarian surgeon, Ignaz Semmelweis, observed a connection between patient health, infectious diseases, and hand washing and other basic hygienic practices. Before that, surgeons washed their hands after surgery, but didn't bother doing so before they operated (or between consecutive operations). They didn't see the point.

In the last fifty years we've been encouraged to think of bacteria as some sort of invisible enemy, one that must be controlled and preferably eradicated for the sake of our family's health. The manufacturers of Lysol must have made a fortune from products that promise to sanitize every crevice of your home, car, or child. Antibacterial gels are ubiquitous. We are waging a full-scale war on an invisible army, though what eludes most germophobes is that by applying chemicals or antibiotics to populations of microbes we kill only a portion of them. Some bacterial individuals have genetic resistance to the compounds we administer. Even if only one in ten thousand individuals survives the first chemical onslaught, it can reproduce quickly to build back the population numbers to replace those who succumbed. The new generation of microbes will be resistant to the chemicals previously used because each individual was born of the ancestor that survived the initial dose of chemical warfare. Hence a so-called superbug is born—one that is resistant to all forms of antibiotic or "disinfectant" compounds.

As our human population increases it is inevitable that we will come into contact with more and more microbial populations. Our historical tendency has been to try and eradicate other populations with the belief that it somehow makes us safer. It has become clear, however, that it is important to maintain a balance with the bacteria whose interests are sometimes at odds with ours. Some small efforts to protect ourselves are preferable for maintaining a healthy balance between "us" and "them," rather than trying to eradicate all bacteria. Physical barriers—such as my water filters—are one way of controlling a nasty microbe population. But modern humans have discovered and manufactured chemical barriers as well. Soap and toothpaste use a substance called surfactant to stop surfaces from clinging to one another; in this case it disrupts the attraction between dirt or bacteria and our skin or teeth. Antibiotics, if they're not overprescribed, can be a useful, manageable shield against bacteria. They work by interfering with the growth and replication of individual bacteria. Today drug companies manufacture these substances synthetically, but in the early part of the twentieth century they were discovered from natural sources.

The Scottish bacteriologist Alexander Fleming was the first person to prove that a fungus called *Penicillium notatum* produced a substance that was toxic to bacteria. Others had made similar demonstrations before him—they were using other fungi of the same genus to show that certain "molds" could prevent bacterial growth. But Fleming formalized the medical treatment of many infections, including syphilis, using these substances produced by the *Penicillium* fungus. He named the wonder substance penicillin in 1929.

Fleming found the antibacterial action of penicillin by accidentally leaving a fungus-laced petri dish too close to a bacterial colony in his laboratory. However, *Penicillium* is not normally found in laboratories.

Instead it lives in forest and field soil, where it forms a weblike structure that is hundreds of micrometers in size. There is a practical reason for *Penicillium*'s deadly touch: It protects the fungi from bacterial parasitizing invaders who might otherwise compete with the fungi for food sources. *Penicillium* feeds by secreting chemicals that break down the carbon found in dead and dying plant and animal matter. It then absorbs the partially digested carbon-rich broth, mixed with water. In this process, a protective penicillin secretion surrounds the fungi colony; this secretion acts biochemically to weaken the bacteria's cell walls and render them helpless against osmotic pressure from surrounding water.[32]

Antibiotics were one of the most profound discoveries of the twentieth century. Treatable infections such as tonsillitis used to be potentially deadly; left untreated they could lead to sepsis and possibly death. However, tonsillitis and other bacterial infections are no longer necessarily lethal; penicillin made them survivable. We have, however, grossly and foolishly overprescribed and misused antibiotics over the last fifty years; some of the bacteria that survive have become superbugs, increasingly resistant to Fleming's great discovery.

Our relationship with antibiotic-resistant bacteria is a case of evolution in action; we can't completely exterminate a disease-causing population of bacteria any more easily than we can eliminate the malaria parasite, or eradicate methane producers from our household pipes. A tiny subset of these populations, either hiding out in undiscovered anonymity, or immune to our counterattacks, can easily rebuild and reinfect us. These "enemies" live on, proliferate, and persist.

Superbugs can be horrifying—such as the flesh-eating MRSA epidemic that is rife in hospitals—and wretchedly life ruining—such as *Neisseria gonorrhoeae,* which no longer responds to any antibiotics in 30 percent of cases; the United States now sees 246,000 drug-resistant gonorrhea infections a year.[33] The CDC estimates that 23,000 people die every year—in the United States alone—of infections from antibiotic-resistant bacteria.

The drug companies are aware of this problem; they have taken this population war and continually tweaked it to an advantage since Fleming's first discovery. We have discovered other organisms, not merely fungi, which also have antibiotic substances. Other bacteria, viral particles, synthetic molecules, and even phagocytic protozoa have been used to fight infections. (A phagocyte is a cell that can eat smaller cells by engulfing them, a process known as "phagocytosis.")

Bacteria can be a political topic too. Municipal drinking water is treated with multiple chemicals in huge holding tanks millions of gallons in size. These chemicals vary from city to city, or town to town. For instance, Ithaca, New York, doesn't add fluoride to the water, whereas in Los Angeles it is added to help prevent tooth decay. The people of Portland, Oregon, recently voted to prevent the city from adding fluoride to its water because most voters saw it as a chemical additive that was essentially polluting their relatively pristine drinking water.[34] It's easy to criticize fluoridation as simply a conspiracy or socialist agenda if you believe that fluoridated water is not the best way to improve dental health. The majority of voters in Portland believed that the city had no right to add medicine or any other ingredient to the drinking water. They thought it was a matter of political ideology instead of public health. But fluoridation—like vaccination—was one of the great civic health initiatives of the twentieth century. Public fluoridation is a well-documented and safe deterrent to the mouth bacteria that cause dental caries.[35] As the humans debate, the bacteria proliferate. In Portland the only "winners" are the bacteria living in the mouths of the city's most impoverished residents, who can't afford to visit a dentist regularly.

I recognize the potential damage that chemicals can cause; however, you have to balance the possible harm against proven benefits. Human beings tend to have very short memories, and it's been one hundred years since the last cholera outbreak in America. But there's a reason our grandparents embraced fluoridation, vaccination, and, in the case of cholera, chlorination; these programs were much needed, and they saved many, many lives. Chlorine should be the least controversial chemical added to municipal water. In most cities you can smell the chlorine very easily by putting your nose next to the stream of water coming out of the faucet. It's not quite as strong as when you visit an indoor swimming pool, but chlorine is readily recognizable. Usually a trip to the country to see your hick cousins makes you aware that there is a difference between the city water you're used to and the stuff that comes from the well. First and foremost, there's no chlorine smell.

The chlorine is added to create a hostile environment for bacteria. It combines with water to form acids that oxidize enzymes and other proteins. This oxidation poisoning kills bacteria quickly. Unfortunately oxidation poisoning is also stressful on human cells, especially those such as skin and respiratory linings, and prolonged exposure can lead to cancers. Chlorinated drinking water systems have been in safe operation for decades, exposure levels are sufficiently low that they don't cause cancer or any other malady. But if you can avoid it, especially through the use of mechanical filters, like those we have in our well-water system, your cells will thank you.

Understanding some basic facts about microbial populations and being aware of their history is the key to sorting out numerous problems of everyday life, from house construction to public health. But having this knowledge—knowing the source of a malady, having an appreciation for the history of the organisms that cause it, and having the tools appropriate for treatment—isn't always the path to a cure. Instead of eradication, it's often more prudent to accept the presence of others and strike a compromise. After all, in any confrontation between microbes and humans the microbes will have a significant upper hand: sheer numbers, the ability to colonize, and the fact that they have persisted since the origin of life, at least 3.8 bya.

The next logical question, then, is, How can these populations be controlled? We may eventually be able to get rid of the methane in our house, but I know that these organisms will not go away, nor do I believe the world would be a better place without them. Bacteria are necessary and essential engines in our environment and, as we will see, also necessary to the proper "education" and function of our immune systems. Even with our whole-house filter in place, I acknowledge that my life is intricately intertwined with microbes. They are far more numerous than one can easily imagine.

Bacteria are ovoid and relatively simple in shape—like tiny sacs. They are diverse metabolically, but they tend to be small, on the order of one to ten microns in diameter. One micron equals one 1000th of a millimeter. The dimension of these organisms is hard to imagine, but one thing that is easy to understand is the need for special optical equipment and staining techniques in order to see them. Bacteria are at least ten times smaller than the average human cell from skin, muscle, or immune system. Because of their tiny size, they are magnificent at hiding in places where chemicals cannot penetrate and larger cells cannot kill them.

It takes a little imagination to picture the world of bacteria. When you are only a few microns wide, a seemingly smooth surface—say a human tooth—becomes a rough and ugly landscape, full of deep fissures and jagged peaks. No matter how diligently you brush, the junctions between mouth surfaces, such as between a tooth and its surrounding soft gum tissue, are almost impossible to keep clean. The spaces are simply too tight and the junctions too deep. It is precisely these places where bacteria find an ideal home and undergo their population growth.

Small size and simple metabolism are keys to the great ecological success of bacteria. But perhaps their greatest "achievement," if you will allow the anthropomorphizing, is their evolutionary longevity. Simply put, bacteria were here first, and likely will be here last as well. These simple colonies of bacteria didn't do much by our standards of accomplishment. Yet they are crucial to life on our planet; they are engines of environmental change. Because the descendants of these long lines are still with us, we can study what they do and imagine the world in which they first appeared.

The fossils of bacteria are called by the same name as their modern descendants, cyanobacteria.[36] They occur as ovoid pairs of cells—only a few microns wide—or longer strands of numerous rectangular cells that may form straight or sheetlike colonies. They can tolerate very high salinity, very low oxygen (they are anaerobic), very high temperatures, and they thrive in intense sunlight. This last point reveals the key to their role as early environmental manipulators.

It's important to understand how microbe populations have changed our environment. They are responsible for creating some of the basic requirements for life on our planet. If bacteria ceased to exist, our fundamental ecological necessities would collapse, and we would quickly die. In fact most familiar life forms would vanish in a matter of days.

For instance, the atmosphere is the ultimate source of nitrogen for all of the biosphere's myriad of species. From bacteria to baleen whale, from slime mold to *Salmonella*, and from millipede to humankind, nitrogen forms the basis of all the protein building blocks for all life on Earth. All proteins are composed of amino acids, all of which have nitrogen compounds as a fundamental component. Roughly 78 percent of the air we breathe is made of nitrogen, but none of this nitrogen exists in a usable form.

Nitrogen is biologically inert. That's why, for instance, it doesn't react with your lung tissue. Unlike oxygen, which gets absorbed by the lung epithelium, nitrogen simply passes into the lungs and passes right back out with every breath.

We build our connective tissues, muscles, enzymes, blood, and virtually all the other tissues of our bodies out of proteins, but we have to get the nitrogen for those proteins by eating other proteins. So what is the ultimate source of nitrogen?

Every molecule of biologically viable nitrogen that enters the biosphere comes from a chemical transformation of atmospheric nitrogen brought about by bacteria. If it were not for these nitrifying bacteria busily converting nitrogen gas (N_2) into ammonia (NH_3), there would be no source of nitrogen for the basic protein and nucleic-acid building blocks of plants, animals, protists, fungi, or bacteria. This process of bacterial activity is known as nitrogen fixation, and it nicely exemplifies the role of bacteria as environmental or ecological engines. If this engine failed, the biosphere would collapse.

What about oxygen? Oxygen comprises roughly 21 percent of the air we breathe. Without it all animals would cease to exist. Humans can stand only a few short minutes of anoxia before severe brain damage sets in, followed by rapid shutdown of all organs and tissues. Unlike nitrogen, oxygen gas is extracted directly from the environment by passing air or water over delicate tissues in all vertebrates. Fishes use gills, terrestrial vertebrates use lungs. In each case oxygen diffuses directly across epithelial tissues in these specialized organs, and we literally steal molecules from the air in each breath. We don't need bacterial help to breathe, but we do need bacteria in order to have something to breathe in the first place!

We often think of plants as the organisms responsible for creating oxygen, through the process of photosynthesis in their leaves. But in fact it is now widely agreed that the photosynthetic organs of plants are the evolutionary descendants of free-living bacteria. All photosynthesis therefore has a bacterial origin. Photosynthesis is a chemical process that converts carbon from CO_2 into sugar and results in the liberation of oxygen (O_2) as a waste product. The rock and fossil records prove that the bacterial engine of oxygen production has been running for billions of years.

Have you ever stopped to think about where the air that you breathe comes from, or what the original building blocks of the biosphere were? It's understandably easy to overlook the fact that our survival is tied into a larger web of life; but only by understanding and accepting the roles of other organisms can we hope to manage the future of our species. The population wars of the past—a series of countless interactions stretching back 3.5 billion years—brought us to this moment.

It is possible that other types of simple cells existed back then, but no trace of them has ever been found. The most conservative estimate is that there were only very primitive biological systems at the dawn of Earth's biosphere. Slightly older sediments do, however, reveal biochemical signatures that indicate photosynthesis was present 3.8 billion years ago.[37] Presumably this was caused by ancestors of the oldest fossils found around 3.5 bya. Therefore geologists are reasonably confident that more than one type of bacterial community was active around the time of the earliest fossils.

Bacterial engines were cranking in the Precambrian, and they haven't stopped since. The sediments reveal these bacterial signatures throughout geologic time, as seen in the Marcellus Shale. The pipes of our houses reveal that those microbial engines are still at work. We can try to stop them with filtration systems, antimicrobial substances, and antibiotics. However, bacteria are so small that we often don't realize that they have us surrounded. Even more foolishly, we concentrate on eradicating them without first acknowledging the fact that we are dependent on them for our own lives.

The human body is composed of so many individual cells that it boggles the mind simply to contemplate such a large population size. Let's say, for the sake of discussion, that ten trillion cells[38] make up a single human being. All of them carry the same genes because they originated from a single original source, a fertilized egg, or zygote. Shortly after fertilization the egg cell divides and begins a long cascade of events known as differentiation and cell division. Cell numbers and cell types proliferate throughout this development.

Various divisions of labor spring up as tissues give rise to organs and the embryo begins to materialize. By the time the individual matures into adulthood, most of its cells come to equilibrium and stop dividing. Some, however, lie in relative dormancy, waiting to be called into action in case of injury or infection, at which time they undergo cell division once more to build up their population size and repair damaged tissue. Some still are in constant turnover throughout life, such as skin, blood, the male gonads, and the alimentary canal.

Ten trillion cells is a huge number. However, recent discoveries show that bacteria—once again—outnumber us, this time on our most literal "home turf," our own bodies. We host, either within or on the surface of our bodies, almost *100 trillion* cells that carry DNA different from our own. All these cells contribute to our well-being and decision making. They are microbes that have formed a unique symbiotic union with our bodies. We all carry an entire ecosystem around within us. And no two ecosystems are the same.

Something is keeping these 100 trillion cells busy; their well-being relies on our personal decisions, and we are only very slowly starting to understand how our seemingly "personal" choices are shaped by the microscopic individuals that live within or on us. We can imagine that our own cells are doing *something* to enhance our lives, even if we don't know exactly what it is. But it is very difficult to acknowledge that ten times more individual cells are working simultaneously alongside them in a microscopic biome—a microbiome—to make us who we are (A biome is a region with characteristic species that is distinct from other regions due to the complexion of its community of species.)

Roughly 99 percent of these cells are anaerobic bacteria that live in the gut. These bacteria break down nutrients that your own cells cannot digest (such as the vegetable product cellulose) and give off gas as a waste product, methane. Your stomach and intestines are full of harsh acids and other fluids that kill most types of living things. Our bacterial partners in digestion, however, are descendants of ancestors that were there when the first rocks were forming on the planet, at a time when only extremely inhospitable environments blanketed Earth, billions of years before any kind of "higher" organisms appeared. Of course they are happy living in the harsh intestinal environment in our guts—it reminds them of "home!"

It's not only cellulose, however, that is being broken down in your gut. Humans are unique among mammals in many ways, but one that stands out is our ability to eat so many varieties of plants and animals. We owe this not to our own specialized cells in our intestines, but to the microbiome composed of bacteria. There are at least one thousand different bacterial "species".[39] Each one of these specializes in releasing particular enzymes that break down various constituents of our highly varied diets.

The genes carried by our gut bacteria determine which digestive enzymes are activated in the digestion of our food. It has been estimated that the microbes in our guts carry 150 times the number of genes that we have in our entire genome.[40] This gives us an enormous metabolic boost and means that there are hundreds of times more products we can digest because of this symbiosis with microbes. We can let the microbes handle the chemical breakdown of foodstuffs, and we can both share the results. The benefits of this symbiotic relationship are huge; our freedom to choose food from many sources makes us highly adaptable. The microbiome is also highly adaptable; it is a community of different species of bacteria that can shift its variety and abundance in response to the food choices of the host. The practical upswing is that if one food source runs out, we can eat another. We can travel beyond the boundaries of a traditional food source with relative confidence that we will find something else to eat.[41]

Bacteria are essential to our health. But we give something to them as well; our intestines are a two-way street. Bacterial individuals are able to acquire genes from host cells and from one another, even though they are a completely different species. This means that many of the bacteria living inside us are carrying some of the enzyme-producing genes of humans. In a sense they are competing with us, using the same enzymes to digest food for themselves that we have eaten. Our body keeps these competitive bacteria in check with bile salts and immunoglobin, two substances released by specialized human cells in the gut.[42] But it's also clear that we benefit from their by-products. For instance, we can digest the sugars that they inadvertently produce through fermentation[43] and we also digest the constant supply of dead bacterial individuals that expire every hour. Each of us is a steward of the other in a very real sense. Our own cells are "irreversibly dependent"[44] on the community of microbes that lives inside our bodies.

Our bodies, and in turn what we consider to be "ourselves," are very much a collaborative effort, and one we have less control over than we like to think. An emerging field in biology focuses on the "gut-brain axis."[45] From studies on other living vertebrates as well as on humans, it is clear that hormonal signals in the gut travel along sensory neurons (brain cells) in the central nervous system to the brain and affect not only foraging behavior (decisions about when and what to eat) but also mood, anxiety, cognition, and pain. Many of these hormonal cues come from the bacterial populations in the gut. This means that some of our most cherished behaviors—being in a good mood, caring about others, working hard, and the like—are not strictly determined by our own sense of autonomy, and therefore they are potentially out of our control. Organisms—ones that carry completely different genes from our own—interfere with our ideas about how to take care of ourselves. Furthermore, it's been shown that dietary changes rapidly affect the makeup of symbiotic species we carry in our guts. This means that vegetarians, for instance, have a different set of bacteria—and equally important, a different set of non-native bacterial genes being activated—than those who eat a diet rich in meat.[46] Think about that for a minute. If different subgroups of humans have different gut flora, due to dietary constraints such as vegetarianism, other differences in culture and lifestyle might be explained by what bacteria they carry.

It seems, then, that we have to see our microbial partners as an extension of ourselves. If we think of ourselves as a huge group of microscopic individual cells living in a shared environment, then what does it mean to be an individual in the first place? If we care about ourselves, we have to protect and serve the intestinal flora as well as our own somatic cells by treating them well. This could be as simple as eating yogurt occasionally for digestive health[47] or refusing to take unnecessary doses of antibiotics because of the damage they cause to all species of bacteria—not just the pathogenic kind, but the helpful partners in our digestive tract as well. It also means, however, having awareness of the invisible population inside us that could cause damage if it grows out of equilibrium.

One classic example is the bacterial gut species called *Helicobacter pylori*. It infects many but not all of us. At low population levels it is a perfectly benign member of the intestinal microbe community. It eats the occasional products of inflammation that occur naturally from time to time along our stomach lining. Usually we can tolerate low levels of inflammation—it occurs naturally whenever we eat certain foods that are difficult to digest, especially industrially produced mass-market foods. In fact *Helicobacter*'s presence in low numbers also promotes those same inflammation episodes. If the inflammation gets out of hand, however, the presence of *Helicobacter* is no longer benign. This microbe was discovered only relatively recently to be the main cause of peptic ulcers, a painful ailment that causes bleeding and degradation of the stomach tissue. And there is a chronic component too. Even the low-level inflammation caused by *H. pylori* can, over ten or twenty years, develop into stomach cancer.[48]

Yet the prevalence of *H. pylori* in so many intestines tells a story that beautifully illustrates how we are inheritors of past circumstances. As it turns out, the presence of this bacterium in our guts prevents other bacteria from taking up residence there. A sort of competitive exclusion is in place: The *H. pylori* excludes other bacteria by secreting deadly toxins. What sorts of "other" bacteria are excluded? The very same species that cause food poisoning in humans. As we saw with sickle-cell disease, there is an evolutionary trade-off to this population war; if you carry *H. pylori* you get perpetual protection every day against getting food poisoning[49] but you also have a very slight chance (around 2 percent over twenty years) of getting stomach cancer. In other words if you can keep this bacterial population at a low level you are healthier than people who've eradicated the bacteria permanently by frequent use of antibiotics. Those who habitually use antibiotics may permanently destroy the low-frequency presence of *H. pylori,* resulting in a higher risk of stomach cancer.

It's easy to see how this phenomenon was produced by evolution. In the past humans didn't live so long. Life expectancy was roughly thirty years in Roman times. That number had increased to merely fifty years at the beginning of the twentieth century.[50] A chronic disease that waits, say, twenty years before it starts to take its toll, such as stomach cancer, would not have impaired anyone throughout most of human evolution because people may have had only twenty years of maturity and reproduction before meeting their early death. Nowadays we have fifty years of maturity (at least) and a much longer time to develop slow, chronic, diseases. Stomach cancer cases have increased because humans are living longer. The longer you live, the more likely chronic diseases—such as those caused by populations with genes other than your own—will bring about your demise.

The presence of a healthy gut microbiota, including *H. pylori*, however, could have greatly improved one's fitness early on in human evolution as brave individuals branched out in their adventurous attempts to try new foods. Those who carried the right combination of bacteria in their guts avoided potentially deadly attacks of food poisoning, and could digest a wider array of plant compounds. It's only now—since so many members of our species are increasing in life expectancy—that we have to deal with the chronic effects of one of those bacterial strains growing out of proportion, such as we see in ulcers caused by *H. pylori*. But treatment in old age is not so simple.

By eradicating *H. pylori* entirely from the gut, with antibiotics for instance, we can cure ulcers. But it appears as though there is a concomitant increase in acid reflux in populations where the incidence of this bacterium is lowest.[51] Over time acid reflux can cause adenocarcinoma, a different type of tumor that affects intestinal and other glands.

By all accounts, then, we have to see the microbes in our guts as coevolutionary partners, meaning that they evolved alongside us throughout the history of our species. They prefer to live inside our guts because it is a suitable habitat, like the ones in which they originally evolved billions of years ago—that is, ones with extreme levels of pH, far away from the deadly activity of both sunlight and oxygen. Their ancestors probably lived in our ancestors' guts, and the entire lineage evolved together. If we view these other species as participants in our evolution, then clearly the common idea that microbes must be destroyed so that we may live antiseptically is utter nonsense.

The narrative of absolute "good vs. bad" doesn't apply here. When organisms interact, they do so because the trajectory of history has brought them into contact. There is no reason to assume that this interplay of species has ultimate winners or losers. Instead we should focus on restoring what has been destroyed through mismanagement or improving historical conditions to make life better for ourselves, which also means maintaining suitable habitats for our coevolutionary partners.

Whether it's the dark, oxygen-free recesses of our home's plumbing, the highly acidic stomach cavity, or the deep-ocean sediment basins full of hydrogen sulfide, bacterial microbes thrive on today's Earth in habitats that have been around since the beginning. Fossils from 3.5 billion years ago, and chemical signatures in Earth's oldest sedimentary rocks, prove that microbial communities existed at that time. Population wars began, it seems, with the very origin of the planet, as soon as appropriate environments had taken shape.

It is hard to know how many species made up those early prokaryotic communities. (All organisms fall under one of two grand categories: Prokaryotes, bacteria, and archaea are unicellular and lack a nucleus. Eukaryotes, the other category, can be unicellular or multicellular, and all have nuclei, among other organelles.) But it is estimated that there are roughly 8.6 million species of plants, animals, fungi, and protists living today, and there are as many as 10 million free-living and symbiotic microbe species.[52] A large portion of these prokaryotes are doing pretty much the same thing they've been doing for billions of years, metabolically speaking. They are ecological engines converting molecules of their surroundings into cellular building blocks. Through all their collective activities they release excretions of gas that alter the atmosphere or surrounding aquatic environment. This mechanistic activity drives the biosphere forward in time as we confront the present day. Each one of us is merely a passenger on board this evolutionary dynamo. But unlike all species that came before, we have the unique ability to imagine the future and manage it to some degree.

5

THE SYMBIOTIC DEPENDENCY OF LIFE, THE VIRAL DIMENSION

I studied and taught comparative anatomy for many years. My students were often surprised that we studied the anatomy of other animals alongside that of humans. A few students would raise their hands and ask why they needed to study sharks (for instance) when it was humans they were interested in. There is a good reason why comparative anatomy requires a consideration of fish (and all other kinds of life) alongside human beings. It turns out that we can trace separate organs, like the liver or immune system, through their own evolutionary history—one that is independent of the creatures that carried them. In a sense the organs can be considered as populations of cells with particular functions. The mechanical workings of specialized organ cells are controlled by genetic information that has been passed down a long line of descent from ancestors who carried similar organs. In this sense organs transcend time and species.

Our organs are ancient; the earliest fishes had livers, and those livers functioned much like the ones we have in our own bodies. No matter what our subject—shark, bird, human—my students quickly realized that each individual is a collection of organs, all of which are related along lines of descent. Liver cells, for instance, function with surprising similarity from species to species. It turns out you *can* compare a shark liver and a human liver and that they do pretty much the same thing.[53] This is a significant realization; it tells us that on a biochemical level these systems have been in place for a long time (just like bacterial photosynthesis). Our bodies are just a collection of these different systems.

I've been thinking about this since my undergraduate training in comparative anatomy, but since that time there has been a growing acceptance of symbiosis in related fields, such as evolutionary biology. In the last chapter we saw how bacteria have formed a symbiotic association with us, living alongside cells of our own digestive system to create a microbiome. This community of different populations with different genomes evolves together and reveals a long biological predisposition for coexistence. In this chapter we will see how a similar type of symbiotic relationship exists also inside us, with a population commonly viewed as "sinister"—viruses.

Microbes such as bacteria and algae are distantly related to us, and understanding their way of life is more an exercise in imagination than in empathy. In general these organisms circulate freely throughout all aquatic environments, and many nonaquatic places as well, moving with the currents and winds, thriving on surfaces that meet their conditions of existence. Inside the tiny fluid-filled spaces between your bones, caked on the insides of pipes and air vents, floating freely in the oceans—colonial and solitary—bacteria are everywhere and their activity has changed little in billions of years.

Even though we cannot visit the primeval Earth or the original habitat of bacteria and other microbes, our imagination has some help in conjuring up their existence. *National Geographic*—both the magazine and the eponymous TV channel—allows us to see up close the remote and microscopic niches that bacteria and algae love. From deep-sea vents in the middle of the Atlantic ocean that spew out poisonous gases and molten magma, to briny intertidal mudflats baking in the sun at 100+ degrees Fahrenheit, to the icy outcrops on glacier-encrusted mountaintops, these Technicolor images reveal the world's loneliest places, where we are shown the evidence of bacterial and algal life. Even if we find it difficult to empathize with the experience of individual bacteria or colonies of algae, we can certainly get a grasp on the inhospitability of their habitats.

Viruses, however, live in an ecological niche that is so alien to us that it's utterly impossible for a human to experience anything like it. Viruses are completely alien to us: Viruses don't make their home in the fluids or watery environments of Earth, or on surfaces or biofilms or anywhere at all that we usually associate with the biosphere.[54] Instead viruses live in the intracellular molecular milieu that only the most specialized scientists have an opportunity to witness.

If recent estimates are correct, there could be as many as 100 million virus varieties currently living among and within us. They are rather simple to describe, but difficult to see because of their size. They are on the scale of only tens to hundreds of nanometers (a nanometer is 1/1000 of a micrometer, which is 1/1000 of a millimeter). Only the most technologically advanced laboratories have equipment that is sensitive enough to view them with electron microscopes. They consist of a capsule—think of it as a very tiny pill capsule—filled with a small number of genes necessary for the replication and synthesis of proteins that pervade the capsule. These proteins are used to bind with host cells. When an infection occurs, viruses attach to the cells of the host. At this point the host cell either ingests the virus completely (a process called phagocytosis; more on this later) or the virus might remain on the outside, but in both cases genetic material is shared; the genes of the virus are fused with the genes of the host in the nucleus of the cell.

Viruses are not self-sufficient replicators; they must make their homes within other cells. Unlike bacteria, which can live their lives free-floating in water because they have all the necessary cellular "machinery" to survive, a virus must find a host cell to complete its life cycle. If the virus fails to find one it becomes—to all intents and purposes—an inert complex molecule that dies without passing along its genes to the next generation. A virus that successfully finds a host cell, however, zeros in on the DNA found inside the host's nucleus. It is here that viral genes fuse with the genome of the host and thereby cause the host's genetic machinery to produce more viral products. After hundreds of clonal viral copies are made, the host cell usually dies and bursts open. This releases the new daughter viruses into the surrounding environment, which begins a new generation of parasitic infection.

Think of millions of infected cells in your nose and throat during a cold or flu, for example, each producing hundreds of new viral particles and then dying every hour, releasing the daughter viruses into your runny nose and saliva. It's easy to understand how the viral parasite population grows so rapidly, and why it so quickly overwhelms your healthy tissue. Sneezing, coughing, or sharing body fluids spreads them to others, where the viral population finds more healthy tissue to infect and continues to increase.

The flu and the common cold are familiar examples of a detrimental viral infection. In other cases, however, viruses do not induce replication in the host cell. Instead they use the host for temporary storage of their genetic material. The host can usually undergo normal operation without hindrance in such cases, even though it has incorporated the viral genes into its own DNA.[55] It's almost as if the virus went only halfway, infecting healthy cells but then resting after it successfully fused its genes into the DNA of the host. Sometimes, perhaps from an environmental trigger, the benign, resting parasitic genes become active again, causing a harmful effect that brings about viral replication. But there are instances where this resting phase has become permanent in host populations. In fact, sometimes it can be beneficial.

One benefit we humans get from viral genes is found in our ability to digest starch. Amylase is an enzyme protein present in saliva. It helps break down carbohydrates in the mouth, converting them to sugars as we chew during the first stage of digestion. The genes that produce amylase are active in the specialized cells of our salivary glands and in our pancreatic tissue. Scientists discovered that a particular viral gene sequence (known as a retroviral one) is embedded in our chromosomes that is responsible for activating the production of amylase in the salivary gland and pancreas.[56] The implications of this finding are profound. They show that at some time during primate evolution, a viral symbiosis took place with our evolutionary ancestors that allowed us to digest starch. Far from being a nasty parasitic infection, this symbiosis turned out to be beneficial because it allowed us to broaden our diet. Such flexibility during times of drought or highly seasonal food availability must have been highly favored by natural selection.

Researchers believe that such beneficial virus-host symbioses are common in the animal kingdom, and that their high selective value allows them to persist for many millions of years.[57] Furthermore, over these vast periods of geologic time, the viral components of genomes, sometimes called the "viral load," may continue to increase because there is no downside to the symbiotic relationship. In other words these viral symbioses are so favorable to natural selection that the viral load can be considered a permanent fixture of many plant and animal groups. Without its high affinity for genetic fusion, a viral particle would be just an innocuous bystander in the pageant of life.

Some scientists claim that since viruses don't have a self-replicating ability they aren't "alive." Calling them nonliving seems to miss a crucial point. They may be unfamiliar to us, but they still have the potential to replicate, and replication is an important factor when we talk about populations. Evolutionarily speaking, for instance, your replicative potential is the only thing you've got that matters. With respect to evolution, your accomplishments, your beliefs, and your ideas mean nothing compared with whether you succeeded in passing on your genes to your offspring and helped them raise successful progeny of their own. We are evolutionarily inert if we don't have children. If you choose not to reproduce, then, as far as evolution is concerned, you cease to have any relevance.[58] Humans may debate about the wisdom and implications of reproduction, but viruses feel no such qualms. Viruses are full of replicative potential themselves, and they provide benefits to their hosts in many cases, so we should consider them an important player in the evolution of all populations.

The year 1964 was a great one as far as I'm concerned. It was the year I was born, though little did I know the magnitude of the globally significant events taking place outside my crib in Madison, Wisconsin. This was the year that the genetic code was finally sequenced after a decade of laborious experimentation. This code is the same for all organisms and viruses. Remember, the "code" is simply the letters and their meaning. The difference between species is not in the code, but rather in the way the code is sequenced. Genetic sequences are written out as a long list of letters, each one corresponding to a particular molecule.[59] The sequence of these molecules, called nucleobases, is different in every species.

After 1964 there was a great deal of interest in finding the DNA sequences of various species to discover just how different each species was from one another. Genetics became a hotbed of research, particularly in efforts to determine if variation at the genetic level (in the sequence of nucleobases) was linearly correlated with phenotypic variation (the varieties of forms visible to the naked eye). This quest led eventually to the development of whole-genome sequencing, which consists of the precise ordering of tens of millions of nucleic acids, and could not have been accomplished without the advent of powerful computers. (The human genome was fully sequenced in 2001.)

Perhaps the most astounding thing we learned about the human genome is the fact that we are not as unique as we like to imagine. The genomic difference between humans and our closest relative, the chimpanzee, is only around 1 percent. That means that roughly 99 percent of our genes are the same as chimps! This would have pleased Charles Darwin greatly. He spent much of his life dealing with people who were outraged by his claim that the great apes and humans came from a common ancestor. People who have a problem with the idea of being related to apes will have an even bigger difficulty with another discovery that came from the deciphering of the human genome.

About 8 percent of the human genome is derived from a particular class of virus mentioned before, called a retrovirus.[60] Another 34 percent mimics viruses that have been inactivated (or possibly just resting, as described earlier), and about another 3 percent has a probable viral origin. Add it all together and almost half our genome is viral. Most of these sequences have been handed down to us from our mammalian forebears. The retroviral component of our DNA alone accounts for a larger portion of our genome than the part that specifies all the proteins necessary for our growth and survival.

The mammalian genome has been under almost constant infection by viruses since the origin of the group more than 200 mya. In that time there have been many instances of benign symbiosis, and as outlined above, in some cases benefits conferred in the coevolutionary process. Another word that describes this fortuitous relationship is "endogenization." We say that a viral genome has become endogenized when it is inserted into the DNA of the host's sperm or eggs.[61] In such an instance reproduction provides the mechanism that allows the infection to pass through time, from one generation to the next. Millions of years later the descendants of the original host are still carrying copies of the original viral infection. Since humans are a late arrival on the planet—our genus, *Homo*, is roughly only 2.3 million years old[62]—we inherited more than 197 million years' worth of viral infections from our mammalian ancestors.

Only about 1.5 percent of our genome is composed of genes that actively produce the proteins involved in building our skeletons, making blood, producing other organs, and all the other things essential to keeping us alive. If you've been keeping count, you see that nearly half of our genome is unaccounted for. Scientists are unsure about its function. Neither active genes nor viral DNA, much of this genetic material may be involved in coordinating the timing of "turning on" genes to make proteins, or "turning off" genes during embryonic development. Some such "control regions" of the human genome are well known already, but there will likely be many more discoveries about how they, and possibly other ones, function.

In order to appreciate viruses, we have to take into account these recent discoveries. In no way do I mean to convince you that we should view viruses with sympathy—to do so would go against my belief that we must view life starkly without regard for sentimentality[63] if our goal is to solve population problems. But first exposure to viruses is usually negative. Each of us has had the flu, probably first acquired when we were very young. The Influenza virus is probably the most familiar of all the viruses. So it's only natural that the public perception of viral coexistence would be negative.

What I've been presenting in this section, however, is less well known, but ever so important to understand because viruses are unique partners in our evolution and in our daily lives, and they have a lot to teach us about coexistence with perceived enemies. Every cell of our body is laden with viral genes. Our cells are of a type known as eukaryotic. Every familiar organism, from plants to animals—even to the lowly paramecium—is composed of eukaryotic cells. The fundamental property that distinguishes these cells from those of prokaryotic cells (that is, bacteria) is that eukaryotic cells contain a complex organization of organelles (tiny organs) while prokaryotes are devoid of them. Each organelle provides a specialized function in the life of a eukaryotic cell.

Bacteria are not so complex. As we saw in the last chapter, they are more one-dimensional in their function. All bacterial cells have an engine-like mechanical function. When this function is active, it can convert chemicals from its surroundings into other chemical compounds and derive some energy for itself in the process. Prokaryotes don't have organelles, they simply carry DNA (not contained in a nucleus) diffusely floating in cytoplasm along with a few other simple parts that provide a metabolic function.

Almost all the life on Earth can be lumped into these two great big categories: those that are prokaryotic, such as bacteria, or those that are eukaryotic, such as animals, plants, or algae. It's taken a long time to figure out that eukaryotic organisms could not exist without the close symbiotic relationship with prokaryotes (for example, bacteria in human guts). This relationship goes back to the earliest days of life; the first eukaryotes were a symbiotic union of two different prokaryotic cells.[64] This union formed a chimera—an organism composed of more than one genetic set of instructions. So if eukaryotic organisms are at root chimeras, then we know that our own bodies are made up of cells carrying genes that were handed down by preexisting ancestral species. Some of these symbiotic unions took place roughly 1.6 billion years ago at the very dawn of eukaryotic life.

Our genetic material is located in the nucleus of each cell in our bodies. You've probably seen a nucleus in a grade-school biology lesson—the teacher cuts a razor-thin sliver of an onion, drops a tiny amount of iodine on it to add color, and the cells that make up the onion become stained. The cell walls and organelle membranes absorb the dye and, once it's your turn to look through the microscope, you see a dark central zone that is itself surrounded by a round darkened membrane, surrounded by a squareish cell wall. The cells abut one another and form a sheet of continuous onion tissue. The circles in each square are nuclei, and the dark-stained centers are the genetic material. It's these nuclei that characterize all eukaryotic cells.

If you graduate to a more detailed course in microbiology, you'll get a chance to look through better scopes and use better dyes. These special stains let you see other ovoid organelles inside eukaryotic cells. These are the mitochondria, much smaller than the nucleus, but much like it in the sense of its origin—their ancestors were free-living bacteria that underwent an endosymbiotic union and now exist as an essential part of a chimeric organism, the cell. It is crucial to recognize this chimeric quality of all eukaryotic cells,[65] and as we will see it involves viruses as well.

Viruses are everywhere, yet they are neither eukaryotic nor prokaryotic in nature. They simply don't qualify as cells. They are, however, replicators, just like prokaryotic and eukaryotic cells. This means that they are able to reproduce and pass on their genetic material to the next generation. Viruses, however, are obligate parasites, which means that they cannot pass on their genes, or make more of themselves, without using another organism's reproductive machinery, as mentioned earlier. This hereditary mechanism gives them an eternal quality: They will be passed along from generation to generation.

Free-living viruses are, to borrow an earlier phrase, evolutionarily inert, unless they find a suitable host to infect. Luckily for viruses they are very good at attaching themselves to a host. The human influenza virus, or "the flu," is spectacularly successful because its particles are expelled when already infected people sneeze or cough. These particles are minute—far too small to see—and so light that they are unaffected by gravity. Instead they float through the air, like sediments in a fast-flowing stream, never touching the ground surface because gravity is not strong enough and their mass is too small to be brought down to rest. When we inhale we suck the viral particles into their preferred oral or nasal habitats, where they attach to the mucous membranes that line our mouth, nose, and lungs, and release enzymes that allow their genetic material to pass into and mingle with the cytoplasm of our own cells. The "body" of the virus,[66] if you wish to call it that—it's really just a capsule—doesn't penetrate the membranous barrier of our nasal, or oral, or pulmonic epithelia. In a sense the viral capsule is sacrificed, expelled as waste by the host, and only the genetic contents are transmitted.

Once the virus's genes are successfully inserted into the host, they quickly attach themselves, by a variety of specific enzymatic reactions, into the already-fully-functioning DNA of the host. In other words, there is no longer any trace of the original viral particle, except in the genome of the infected organism that now carries the viral DNA. The infected person quickly starts to feel the effects of the flu because her cells are not functioning normally. Her metabolism is now impaired by the encumbrance of the viral load. The virus uses the human's cellular mechanisms to make copies of its genes inside the infected cells, to build new capsules that carry the reproduced viral DNA, and to cause cellular activities that are not part of the host's usual healthy functioning. The result? In the case of the flu, the infected person gets really, really sick. The virus needs bodily fluids—snot, phlegm, or diarrhea—to leave the body and find another host. The host's malfunctioning organs and membranes oblige, generating streams of fluid. This makes her life wretched but allows the virus to leave in search of new fertile habitats—that is, more uninfected bodies.

The departing virus takes a souvenir of its host with it; most virus species have a membrane (aka the envelope) that surrounds the capsule (aka the capsid). This membrane is made of a bilayer of glycoprotein and lipids. When the virally infected host cell produces new viral particles, a fresh double layer of lipoproteins is created out of the membrane-building machinery of the host. In other words the offspring viral DNA[67] is enclosed by membranes made by the host. Two different organismic sources (host and virus) produce one chimeric offspring.

In the life cycle of a typical virus-host interaction, the end result is a co-opting of the host's ability to make copies of itself and produce cell products. Viral DNA gets incorporated into that of the host, and unwittingly the infected cells begin to make viral copies in the process of cell division, reproduction, and normal metabolism. This co-opting of cellular machinery can result in the budding of membranes from the infected cells—membranes that are produced by the host. Viral DNA can direct the actions of host cells to produce new membranes, surround copies of viral DNA inside the cell, and create buds that are ejected and sent out into the extracellular spaces. When viral buds are released, they become virions, or viruses in search of another host to infect.

The moment an appropriate cell surface is encountered, such as your nasal epithelium, for example, a virion attaches, releases its genetic material into the cytoplasm of the hapless host cell, and the process of viral co-option begins again. The host cell bears the infection by carrying the burden of viral genetic material. There is no competition, no war, just an automatic symbiosis whenever virus meets host.

Evolution is a series of intricate and complex interactions among living things, and viruses' behavior is a great example of this. They nicely reveal one of the recurring themes of this book: All organisms are living proof of past symbiotic interactions and are simultaneously being influenced by new biotic interactions that will change them in ways we can't necessarily anticipate.

At some point in the distant evolutionary past, certain infections seem to have become permanent fixtures of the biosphere, such as the beneficial viral infections mentioned earlier, or as we will see, the photosynthetic organs of all plants. While viruses themselves are good examples of symbiotic unions (the capsid membranes are produced by the host, and by genetic material from the "parent" virus), eukaryotic cells show an even richer history of symbiosis, one in which viruses also play a role.

Plants and animals are familiar to us, but their cellular anatomy is complex. If we look carefully at two basic types of eukaryotic cells, a generalized plant cell and a generalized animal cell, we see that both have significant differences, even though both are eukaryotic. These differences include that plant cells have a rigid cell wall made of cellulose carbohydrate while animal cells lack a cell wall; plant cells have pigments that absorb sunlight, while animal cells lack such pigments. But the most important difference for our discussion is that plants derive their energy from tiny organelles called plastids (chloroplasts, for example) while animal cells lack them and instead derive energy from organelles called mitochondria.

One similarity between plant and animal cells is the presence of a nucleus. It is inside this—the largest of all organelles—that the genetic material, DNA, is stored and sequestered when the cell is in its resting (nondividing) stage. The nucleus, mitochondria, and plastids of a eukaryote tell a fascinating story of chimeric evolution.

The eukaryotic cell is composed of more than one genome. That is to say the genes that it carries in its nucleus have a different evolutionary history than do the genes it carries in its mitochondria or plastids. The Russian biologist Dmitry Merezchkowsky first suggested, in 1905, that the eukaryotes had essentially evolved by an ancient symbiotic event so successful that it gave rise to all the various forms of eukaryotic organisms alive today. The late professor Lynn Margulis, of the University of Massachusetts at Amherst, revived this notion (having discovered it independently) based on more modern data in the 1980s, emphasizing the similarity between mitochondria and certain archaebacteria. In the 1990s James Lake at UCLA advanced the notion that the nucleus itself was the product of an ancient endosymbiosis, having originally been a free-living archaebacterium.[68]

Today there is little argument among microbiologists that endosymbiosis—the engulfment of bacteria by other bacteria—played a key role in the evolution of the eukaryotic cell. In fact it is referred to as the "endosymbiotic theory" of evolution, and it takes up a good deal of one lecture in the introductory evolution course I teach. There is considerable disagreement, however, as to which particular bacterial species were involved in forming the nucleus, mitochondria, and plastids.[69]

For now it is safe to say that the nucleus, mitochondria, and plastids of all eukaryotes each carry a different set of genes, which are leftover signatures of ancient symbiotic events.

When a cell divides, as happens over and over again during our growth, the nucleus makes copies of its genes, and they are distributed to daughter cells so that every cell in our bodies contains the same genetic material that was present at conception. There is no sharing of this genetic material with the mitochondria (or plastids in plants). The mitochondria that were present at fertilization (provided by the egg, i.e., from the mother) divide independently and contain their own suite of unique genes—originally from a certain primitive species of bacteria that belongs to the group called proteobacteria—and do not contribute anything to the nuclear genome of the host cells. The same is true in plants. The nucleus contributes its genetic contents only to the nuclei of daughter cells, whereas the plastids contribute to daughter plastids, and never do the two suites of genes commingle.[70]

So if eukaryotes are the product of a symbiotic event between two different prokaryotic organisms, this leaves us with a couple of puzzling questions with respect to this chapter: (1) Where do viruses, which are neither prokaryotes nor eukaryotes, come into play in the endosymbiotic evolution of eukaryotes? And (2) How does this relatively new, symbiotic worldview mesh with the more traditional doctrine of competition-based evolution?

We will leave this second question for another chapter.

But to answer the first question: There are something like 100 million species of virus on our planet, and many if not most of them infect only bacteria. As mentioned earlier, these viruses are known as bacteriophages. A bacteriophage virus injects its genetic material into a bacterium, and the genes of the host are mingled with the genes of the virus. The cellular machinery kicks out copies of the viral DNA and packages it into tidy little capsules that are ejected from the cell (often at the expense of the cell; it succumbs to the infection), and the new viral particles (daughter virions), part viral (genes), part host (membranes), are released to an unsuspecting population in search of more infectious activity.

The fusion of viral DNA and host DNA led microbiologists to consider a possible role for viruses in the endosymbiotic theory of evolution. If bacteria were engulfing other bacteria in the Precambrian, forming endosymbiotic unions, leading to a more complex type of cell (eukaryotic), this might have been a good defense against bacteriophages. In other words, what better way to protect yourself from viral attack—which occurs following the virus's "recognition" of proteins on the cell membrane surface of the host—than by taking refuge inside a different species?

Viruses filled the Precambrian oceans, looking for potential hosts to infect. This still happens; free-floating DNA viruses (called mimiviruses) roam our oceans, and like bacteria, they have probably been floating in all kinds of watery environments since the earliest phase in the history of the biosphere.

Perhaps it was in the open ocean that another unlikely but highly adaptive event occurred: Free-floating viral DNA became incorporated into the cytoplasmic space of another free-floating unicellular organism, something akin to a red alga. In other words the origin of the eukaryotic nucleus was, in simplistic terms, a virus escaping from a free-floating life stage where it was under constant infection by other viruses—to a coddled existence inside another cell (such as a red alga).[71]

The nucleus itself, then—the defining characteristic of all familiar organisms[72]—can therefore be thought to have a viral origin. Consider these facts: Viruses can disintegrate the membrane of other cells and reassemble it, just as we see in the nucleus during eukaryotic cell divisions. The nuclear membrane disappears during metaphase and is reassembled during telophase in the growth and maintenance of all familiar organisms. This implies that viral genes are involved in genetic replication during cell division of eukaryotes—a fact that has been confirmed by molecular biologists.

Perhaps most intriguingly, viral genes are directly involved in the transposition of eukaryotic genes during cell divisions. In other words viruses assist "jumping genes" in eukaryotes. This phenomenon—in which segments of DNA get transposed from one chromosome to another—is one of the most significant sources of genetic variation in familiar organisms. And as we already know, evolution is not possible without genetic variation. If you look at life this way then viruses are crucial to all eukaryotic organisms.[73]

Evolution of familiar species including humans involves viruses from the Precambrian oceans. Today's viral diseases can be seen as more recent phenomena, patterns of coexistence that are still being worked out. The ancient infections have evolved over hundreds of millions of years and are now part of us, sequestered in our chromosomes. They can now be considered symbioses.

Some people find the idea that viruses played—and continue to play—a part in the ongoing evolution of life distasteful. Who wants to be related to the flu? These people might argue that prokaryotes—bacteria, for example—eventually evolved into eukaryotes and that viruses were equally infectious to both. There is evidence, however, that viruses were more benignly intimate with eukaryotes. Let's look at the enzymes that help repair and replicate genetic material, as well as the protein that forms the "skeleton" inside eukaryotic cells (a protein called tubulin). The polymerase molecules, DNA polymerase and RNA polymerase, as well as the protein called tubulin, are crucial to the proper functioning of all eukaryotes. The polymerase enzymes work to assemble new chains of DNA and RNA during cell division, and the tubulin protein acts as a scaffolding upon which chromosomes assemble and get pulled apart during the formation of daughter cells during replication. These elements of eukaryotic cell division can be found in viruses, but prokaryotes have their own rudimentary polymerases—unrelated to those found in viruses and eukaryotes. Tubulin is not found in prokaryotes at all.

Given this information,[74] it is unlikely that prokaryotes "evolved into" eukaryotes during the early days of cellular life on the planet. However, it's equally unlikely that viruses "started the evolution of life," because they are completely dependent on other life for their existence. The only logical conclusion is that all organisms today have a symbiotic dependency on other life forms due to the interactions of the simplest single cells, free-floating viruses, and organic molecules (building blocks) of the primeval oceans. Those relationships that were formed at the earliest phase of the biosphere are essential to the functioning of the biosphere. We cannot escape these interactions. The vestiges of those ancient symbioses are readily observable in ourselves and in all organisms with whom we coexist.

The origin of the eukaryotic cell is most likely a three-part drama that took place hundreds of millions of years before any kind of complex life arose on the planet. First came the endosymbiosis of mitochondria (proteobacteria infecting another unicellular organism), then the endosymbiosis of plastids in plants and protists (cyanobacteria infecting cells that had already evolved with enclosed mitochondria), and finally the endosymbiosis of the nucleus (a large mimivirus-like virus infecting an alga with plastids and perhaps a separate infection of plastidless cells with mitochondria). All three acts were symbiotic in their style. This means no competition, no warring factions, no individuals maintaining a dominance over other individuals, but rather the automatic molecular happenstance of polymerized organic molecules (DNA, RNA, and their associated products, enzymes, and the like) coming into contact with other molecules to form more complex life-forms (such as rudimentary cells). Symbiosis set the stage for all that came later, colonial cells, tissues, organs, and multicellular organisms.

All molecular and genetic evidence points to a counterintuitive conclusion: It was not some primitive prokaryote that exhibited particular favorable variations and evolved into a eukaryotic organism. Rather it was a series of favorable symbiotic events that allowed a large virus to infect some sort of primitive alga and give rise to a lineage of organisms that forever contained a membrane-bound nucleus.

Billions of years and countless infections later, these viral organisms are an integral, endogenous part of our own species' genome. What are they doing there? The same thing viruses have done since the earliest stages of Earth history: inserting themselves into other organisms' chromosomes and using the cellular replicative machinery of host organs to make more copies of themselves.

The astonishingly high viral component of the human genome[75] reminds us that the genetic variation we all carry is due not only to the inheritance from human-like ancestors living on the savannah in the geologically recent past, but also from parasitism and modification by viruses, in symbiotic unions with more ancient roots.

As alluring as it may seem, the traditional Darwinian concept of "survival of the fittest"[76] does not hold up in the light of what we know about endosymbiotic theory. Since viruses have been around since the Precambrian, constantly looking for new cellular hosts to infect, it should come as no surprise that they can jump from one species to another (as in the case of HIV, avian flu, swine flu, SARS, and the like). When they make the successful jump—a kind of intrepid exploitation of a new environmental niche—they go on to modify the genome of the species they infect. In this respect they are not the kind of gene transmission we have been taught to think of as "adaptive." In other words it's not always the "more fit" genome that gets passed on to the next generation. It's also the lucky ones that happen to engage in a beneficial symbiotic union. "Fitness," the ability to find a mate and leave many descendants, plays less of a role, and good fortune, and the ability to coexist with other living things, plays the bigger role in this view of evolution.

This means that the traditional Darwinian understanding of evolution—as individuals locked in a vicious "struggle for existence" in their adaptive quest to meet the demands of their environment—is not adequate. An individual is simultaneously coping with the circumstances of its own existence—the genes it inherited from parents, the bacterial load it carries on itself and inside its digestive, alimentary, reproductive, and excretory canals, and the viral load that is living and operating within its cells. All these factors together offer variables, in addition to the variables of the external environment, that affect the health and well-being of each individual in a population. The abiotic environment is perhaps the least important of these. The symbiotic environment may be the most.

Furthermore, natural selection functions not simply with respect to the heritable DNA of the species' own gametes, but rather on a more inclusive set of genes from multiple organisms: the "host," the bacterial flora in the gut, and the viruses inserted into the host's genome. This tripartite collection is known as a holobiont, and since each component is dependent upon the others, evolution proceeds only with respect to the entire holobiontic organism. All eukaryotic organisms function in this interactive and symbiotic way, which means that adaptation is essentially a cooperative, rather than a selfish, enterprise.

Richard Dawkins popularized the idea of the "selfish" gene, expanding on the work of evolutionary theorist William Hamilton, as a way to envision how altruistic individuals evolve. Individuals come and go. They can be thought of as mere vessels that carry the real stuff of evolution, the genes that are passed on to offspring. The Dawkins worldview sees the genetic instructions that we carry as directives to do things that are in the genes' best interest. Sometimes altruistic behavior isn't in the best interest of the individual displaying it. Think of running into a burning house to save your three children. That act, which almost any parent would do at a moment's notice, might result in an early death for the rescuer, but if it saved the kids, it would be worth it. From the selfish gene's perspective, even if the rescuer died in the act, as long as the kids survived, three copies of 50 percent of the rescuer's genes would live on. In any group of closely related individuals, self-sacrifice might actually promote the success of a large number of relatives who also share the altruist's genes. Hence the gene benefits selfishly to make sure more copies of it are replicated, regardless of the health and well-being of the person carrying it.

Although it is often used to understand altruism—and is still very controversial at that—the "selfish gene" concept fails to explain the symbiotic tendencies of most life-forms. While we can see that viruses as well as bacteria may have initial stages of "plague culling" (high mortality rates during early phases of contact with new hosts), this is soon followed by coevolution of microbe and its host.[77] Instead of replicators acting selfishly, it seems there is a stronger tendency for organisms to evolve together.

All this information leads us to one thought: that humans are more than just the sum of their DNA and their own adaptive "design." Instead each of us is a holobiontic union of genetic material from mammalian ancestors, bacterial "machines," and viral infections. All these genetic components work in concert to make us who we are. In one sense the union we form is a mingling of organisms that came together by happenstance. But it is crucial to understand that the holobiont is subject to natural selection, and that the different genetic systems have been fine-tuned over the course of evolution.

Perhaps the best illustration of this is where we started off this chapter, the endogenous retroviruses and their relatives that make up nearly half of the human genome. If this genetic material were simply "baggage" or "junk" DNA, we could liken it to a viral load on a PC, but this analogy is not sound. Everyone who uses Windows PCs remembers not so long ago, before virus-blocking software became standard equipment on new computers, that every time a computer made contact with the Internet it became the target of small indiscriminate programs that corrupted the boot-up process or the launching of a favorite program. These "computer viruses" are aptly named because they insert themselves into the informational stream of a program and corrupt its normal functioning. Computer viruses never led to any kind of evolutionary innovation in the workings of the computer programs, and this is because they don't participate in the creation of new computer programs. They are released by unscrupulous hackers only after new programs are launched as the new "generation" of products hits the market. In this respect computer viruses *are* a decent analogy—in the sense that they insert themselves and co-opt the hardware of the machine by interfering with the host's software—but since they don't have any way of getting passed to new generations, they cannot be thought of as symbiotic.

We have a symbiotic relationship with viruses; we can't live without them, even if we wanted to. This is a hard message to sell to a germ-obsessed public, but it's the simple truth and it explains how our genome came to be riddled with so many viruses or viral components.[78] Our relationship with viruses is complex; a recent discovery demonstrates that certain enzymes, found originally in viruses but lacking in bacteria, are active in determining the developmental path of human and indeed all animal embryos.[79] Without this enzyme (reverse transcriptase), embryos fail to develop beyond the first few stages of life. Interestingly, reverse transcriptase was also found to be abundant in cancer tumors. Indeed this discovery helped point the way to a vital function for a large portion of the human genome whose function was previously unknown. Somewhere within this mass of genetic information that came from our viral symbionts lie the switches and signals that control the timing of cell division, and specify the shape and function of developing cells.

It's not surprising that reverse transcriptase is most active in embryos and tumors. Both are comprised of cells that replicate prolifically, at undifferentiated stages in their life cycle. The chemical tendency of viruses has always been rapid and prolific replication, all the way back to a time when life was simpler, before the advent of eukaryotic cells, deep in the Precambrian past. The almost miraculous formation of an embryo and the grotesque and formless mass of tissue called a tumor are two sides of the same coin. In both instances reverse transcriptase is at work, dispassionately guiding the proliferation of eukaryotic cells[80] to one end or the other.

The viral genomes that exist within us began as infecting agents and long ago killed off those individuals who were most susceptible to their pathological effects. After endogenization, they accumulated in the nuclear DNA of our ancestors. We share an important tradition with all plant and animal species: Viruses have acted like freeloading copilots inside us, hitchhiking and partially directing the ride through evolutionary time.

6

ESTABLISHING A WAR NARRATIVE FOR POPULATIONS, THE IMMUNE SYSTEM

Most summers are tour season for Bad Religion. It's a familiar routine; I'll make multiple international trips from May to September. Usually I fly back and forth to see my family at least a few times every month. I know which flights work best for me and which airplane seats to reserve. I can guess at the quality of the service and the potential frustration factor of the process, but one thing I can't predict is the behavior of the other passengers. One of my biggest concerns is staying fit and healthy through the arduous tour process. However, on the first flight of our 2014 tour I was reminded that all the healthy eating, workouts, and nights of good sleep can't keep me safe from all contagion.

Two seats away from me a young woman started sneezing. I was a helpless passenger—already buckled into my seat—and I knew there was no escaping an imminent deluge of sneeze-spray into the recycled atmosphere. Soon the Kleenex was out, and an interminable cycle of nose blowing and sniffling started. There was nothing much to do except try to slow my breathing and exhale more than I inhale. I have no idea if this actually works. Even though I couldn't see them, I knew that I was already surrounded by microscopic aerosol droplets filled with antigens ejected from a stranger's mucous membranes. In this case the DNA would be from viral or bacterial spores spewing from the woman in seat 6A.

We sample populations of bacteria and virions with every breath we take. However, the bacteria and viruses we inhale usually don't make us sick. This is because of an impressively elaborate organ called the immune system. The workings of the immune system are highly instructive to our discussion about populations; the immune system is constantly responding to changing environmental conditions while engaging in an evolutionary process to select antibodies against invading populations.

The most common kinds of airborne microbes in the most-often-visited environments have been sampled many times, usually very early in our lives. Our immune system has been "educated" by this exposure. Since it has encountered these populations of microbes before, it has built up a suitable resistance to them. Whenever we came into contact with the most common microbes, they were destroyed by the specialized array of cells and molecules in the blood of our own bodies and circulated in the lymph and plasma of our vascular networks. During my first flight of the year, I couldn't help fixating on the possibility that this might be that day when I came into contact with a virus or bacterium that was totally new to my immune system.

Soon it was time for the "meal," and a stewardess started handing out fruit plates with cheese. I knew that those sad strawberries and melon chunks had been manually prepared not too long ago and had probably sat in a holding dish before they were finally sorted into the hundreds of portion-size trays in which they were now being distributed. Did I dare take a bite? Airborne aerosols aren't the only place microbial spores exist. They also populate surfaces. Bacteria love human food, and anything that is smeared with organic matter. Who knew if the food preparers remembered to wash their hands before stacking the plates? For all I knew one of them might have sneezed on the cantaloupe. I'm no germophobe, however, and since I was hungry, I didn't hesitate to eat. I had faith that if there were any microbes present in the food, their antigens would be recognized by my immune system. I've been on hundreds of flights, mostly with the same airlines. My immune cells have come into contact with similar bacteria and viruses before, most likely, and they would recognize and kill them without causing any kind of noticeable reaction.

But there have been times when I don't feel such faith, because I know that immune systems are fallible. We are particularly vulnerable in unfamiliar environments, where our immunity cells won't recognize the local microbes. Of course global travel means that these unfamiliar microbes can take their own flight right to our hometown, where they can rapidly spread among an unprotected population. We are even vulnerable to new domestic strains in our own backyards. This is because of mutation. Microbes (and, as we shall see, the cells of our immune system as well) are notorious for rapidly shuffling their DNA.[81] Familiar environments aren't necessarily safe; a new mutant population of bacteria or virus might show up at any time. This is why my spirit always sinks when someone near me is sneezing. It could be some new mutant strain of flu or cold, since the organisms that cause these maladies mutate so readily.

As I sat on the flight I was aware that my cells might soon be engaged in a true population war, one that could easily have been prevented if the woman sitting in my row had just covered her mouth when she sneezed and washed her hands on a regular basis. Alas, my neighbor seemed untroubled by any concern for her fellow passengers, carelessly spraying the cabin with an unknown mixture of eager bacteria and viruses throughout the ensuing flight.

I use this example as a demonstration of everyday situations we find ourselves engaged in, unaware of the population wars going on simultaneously. Even though I can't feel the biochemical reactions taking place in my bloodstream as pathogens are destroyed, the end result—waking up the next day feeling healthy—makes me feel grateful for the unconscious protection afforded by my immune system. The immune system is a population of cells undergoing evolution (descent with modification spurred on by selection). Sure, it is an organ—not a glandular one like the liver, but it's one whose cells change drastically over the course of only a few hours or days at a time. This dynamic organ, which we usually consider as a kind of "defense mechanism," is illuminated particularly well by the metaphor of battle. But that metaphor extends only so far, as we will see. The immune system is not in any way hell-bent on world domination, as most conventional armies appear to be. Furthermore, like all populations, the immune system depends on coexistence and equilibrium with pathogens. In other words, in some sense the "enemy" is necessary for a healthy immune system.

Two weeks after arriving in Europe, with no evidence that I had contracted any sickness on my flight, I visit the *Dom* (cathedral) in Cologne, Germany, the most important Gothic building north of the Alps, where Catholicism exerts its mysterious attraction on believers and pilgrims of all nationalities. I may not "believe," but I can respect the human achievement involved in the building of it, especially considering that work started in 1248, centuries before the first bulldozer or high-rise crane. Upon entering the massive stone-walled structure, my eyes have to adjust because lighting the cathedral's nave, a space with 142-foot ceilings, is nearly impossible. The huge stained-glass windows refract a glow of sun, which is filtered through with dark reds and blue hues that seem to barely illuminate the endless rows of empty pews. Hardly anyone is sitting in them, but the place is populated by observers. People of all nations, it seems, wandering aimlessly with their necks kinked to view the craftsmanship and eerie ambience.

I stand next to a teenager speaking Spanish; he's wearing cutoff shorts and a T-shirt that says "Hellfest 2013." We are joined by a noisy gang of Americans, talking loudly about their visit to the Cathedral of the Saint John the Divine in New York last Christmas for a new-age music concert performed by Paul Winter. In fact, it seems that most of the visitors from distant lands were simply eager to marvel at the grandiosity of the building. I noticed, furthermore, that only a very small percentage of people who enter this "active place of worship" seem interested in any kind of traditional prayer or pilgrimage.

Despite the ample confession booths and availability of candles for devotionals, only a couple of people are on their knees to pray or meditate. Most people seem to be inspired by the architectural grandeur, regardless of their cultural background. I see Chinese and other Asian tour groups being led around by interpreters carrying little flags depicting various nationalities. I hear various non-European languages being spoken by other tourists—Russian, Hindi, American English, and French Canadian. Many of the younger visitors are listening to their iPods while they amble along the rows of pews toward the empty altar. Both the altar and the bored teenagers look lonely—the former appearing strangely underutilized, and the latter too young to understand why their parents dragged them to this tourist trap. This contrast of modernity with the unchanged formality of Catholicism makes them both seem out of place. The world has changed drastically in the millennium since the Dom's construction began.

When I began coming to Europe in 1989, the Dom, like many other European churches, staffed doormen who would not allow entrance to the sacred church unless appropriate clothing was worn and a proper attitude was exhibited. A youngster wearing shorts or a woman showing too much skin would be turned away. Likewise, if someone was listening to headphones, they would not be allowed inside. Loud talking and laughter were discouraged because, as the signs read: "This is a place of worship, please be respectful." Then as now the church collected an entrance fee—money that went into the general fund to maintain the grounds and to restore the building itself. One hundred years ago visitors could enter only if they attended mass, were very formally dressed, or paid some sort of fee (a tithe, perhaps) that was expressly for religious purposes.

What happened to Catholicism? Why would the Dom allow such a diverse array of the disrespectful and non-Christians inside its walls today when it was built expressly for the pious worship of Jesus and his father? Did the Catholic Church become more forgiving, more liberal, and more modern? Or has the appeal of the secular aspects of the Dom (the architecture, the social gathering place) become co-opted by modern society? I think it's the latter. The Dom is the most crowded place in town not because Catholicism has become more popular, but because secularism has taken over the role religion held in the past. Meanwhile, Catholicism in Germany is largely a social formality without many pious devotees. German households have the lowest percentage of church attendance in the entire Western world. But as the Dom demonstrates, the Germans take their preservation of culture and architectural heritage very seriously. The city officials and Catholic leaders in Cologne recognize the importance of the Dom to tourism, and as a central icon of the city. These things are far more important in the modern context than the religious significance of the building. The tourist entry fees have gone up, creating more revenue for structural restoration of the church and for staff employment. Loosening the rules of who can visit has proved good for business.

Cologne can't afford to have its biggest—and, after the bombing raids of WWII, arguably only significant—tourist attraction turn away visitors. It's good PR for the city to popularize such an important Gothic building. The Dom now has more than one purpose: It needs to fulfill its original religious purpose as well as appeal to modern secular interests. The vast and imposing Gothic cathedral hasn't changed, but the needs of the people who visit and manage it are in flux.

The Dom is one of my favorite places in Cologne. I can never escape, however, from its constant reminder that clinging too tightly to a rigid teleology gets us nowhere. Like all things in the universe, its purpose is always in flux. Nothing stays the same forever. Even something so seemingly purpose-built as a cathedral changes its function and meaning over time. In this respect we can analogize the Dom with the organs and tissues that make up the vertebrate body.[82] The structures—the arches, buttresses, spandrels, cut stones of the foundation, and so on—haven't changed significantly throughout the life of the building. They have been reshaped now and then—such as when some of the frameworks and stained-glass windows were destroyed in WWII and subsequently rebuilt with new materials, altering their character slightly, perhaps, but generally appearing and functioning the same as the originals. Throughout the history of this structure, however, the people who use the building have changed quite drastically.

Populations change through time, but their organs and tissues are inherited remnants from their ancestors, and remain relatively permanent throughout the evolutionary process. We all use the architecture of our ancestors—the body plans, organ systems, and reproductive methods—and they are shaped, modified, and sometimes repurposed for our own particular modern-day needs.

In one sense organs and tissues from our ancestors can be thought of as a template upon which to build new structures. In that respect we recognize that there are historical constraints to what is possible. Lungs, for instance, evolved from an organ in fishes. They didn't arise de novo from some miracle; rather, lungs developed slowly over millions of years, deriving from swim-bladders in fishes hundreds of millions of years ago. Countless vertebrate species have used lungs ever since. Could a better lung be engineered? Probably. Some species, such as birds that live above twelve thousand feet on Mount Everest, have exceptional lung features that allow them to extract oxygen from the hypoxic high-altitude air. Their adaptations include a more efficient style of breathing, allowing countercurrent flow between air and blood in the lung capillaries, higher surface area of the lung tissue, and specializations for greater capacity of oxygen diffusion from blood to the mitochondria inside muscles.[83]

In all vertebrates, lungs extract oxygen from air and expel CO_2. Why don't we wear our lungs outside, or maximize the surface area for gas exchange all over our skin? The answer is because of the developmental constraints that all vertebrates (including humans) inherited from our fishlike ancestors. Lungs will always develop inside the body cavity, always be connected to the atmosphere via a narrow opening (the trachea), and therefore always be limited by those factors. This configuration has persisted for more than 390 million years, and not a single fossil has been found that shows a different functional arrangement for lung breathing. If evolution results in a different configuration sometime in the future, it will be a true novelty. All the known varieties of vertebrate species that breathe air have lungs, some of which have been modified to some degree, but they all have essentially the same structure.

As strange as it may sound, this is what goes through my mind whenever I visit my favorite cathedral in Europe. The architectural structure has remained as a monument through time, a gift from our ancestors, but its patronage has changed drastically, and therefore the building has taken on a new meaning to the current demographic. We can see the way its population (the collection of visitors sampled at any given instant in the long life of the cathedral) has changed over the years. Now, when I put on my evolutionary biologist hat, I consider the significance of that change.

Some would claim that Catholicism has lost its significance in modern society, hence the demographic change toward a more secular patronage. Others, however, would counter that Catholicism is just as important as ever, as evidenced by the throng of people interested in visiting the Dom. They might concede, however, that the religion has changed its tolerance and practices to appear more "modern" and attract new souls into the fold of Catholicism.

One of the central themes of this book is that population "wars" result in assimilation rather than obliteration. Catholicism hasn't been destroyed. It seems to coexist peaceably within a highly secular society. In terms of cultures, history often portrays a narrative of vanquished populations at the expense of the victors in classical warfare. But biology teaches us something different: Ancient cells and organs—like European cathedrals now serving other purposes—are what modern species use to get by. Catholicism has changed and must continue to change to endure. As I write this the Catholic world is in uproar over an interview given by the current pope. He acknowledged the unthinkable: The Catholic Church needs to get over its fixation on abortion, birth control, and gay marriage if it wishes to survive. In other words it needs to adapt to the realities of the modern world or risk losing even more souls of the next generation.

Evolution doesn't invent new cells or organs very often. In the same sense, once organ systems have been established by natural selection, they don't go extinct (though some organs lose their function—for instance the human appendix, which was originally larger in our ancestors, as seen in other mammals, and used to digest cellulose at an earlier stage of mammalian evolution). Through the long course of evolution, organs have retained their physiological functions, even if sometimes they get used in new ways. It's not at all uncommon to find ancient organs co-opted, or perhaps "improved upon" by more recent taxa, while at the same time retaining their basic functions under new environmental circumstances. Nowhere is this better illustrated than in the immune system.

As mentioned above, the immune system's function typifies the classic tale of victory and defeat in warfare. Most of us know that germs, microbes, and cellular invaders can cause disease. We don't always fully understand how they work or where these organisms originate (there are just as many "germs" on the silverware and tables at a restaurant as there are on the toilet seat or bathroom door handle, for instance), but most people think that a total annihilation of all germs is not only desirable but attainable. Therefore, when we speak of the immune system as a cellular protection system for our healthy bodies, it's not a far stretch to explain it as a victorious conquest over the ne'er-do-well germs. To most people total obliteration of any invading microbe is the "goal" of the immune system. Indeed this is how medical textbooks explain the workings of immunity.[84]

However, it is crucial to understand what the immune system actually does. The immune system's job is to maintain a germ-free environment for an individual human body; this does not mean that the population of a microbe species is destroyed, only that a portion of it has failed to find a suitable environment (that is, your body) in which to proliferate. The germ population will continue searching for fertile ground in other people (for instance, humans who are immune compromised or genetically susceptible) or other species (for instance, viruses that travel between species, like swine flu or SARS). We can only consider a germ or pathogen[85] to be vanquished when all subsets of the host species are equally free and clear of infection. And this is a very rare thing.

Plagues, flus, and exotic outbreaks have tormented human beings for millennia. The bubonic plague (also called the Black Death) was one of the most destructive maladies to ravage human civilization. Its victims died a slow, painful death, and medieval doctors had no way to understand or cure the disease. In the 13th century, twenty million people in Europe and the Middle East succumbed to its ravages. By the late Middle Ages, bubonic plague was responsible for wiping out nearly 50 percent of the standing human population.[86] Even in the twentieth century tens of thousands succumbed to plague, mostly in Africa, but also about one thousand people in the Western United States. Cases in the United States continue to pop up every year, and internationally they are more numerous (more than one thousand cases per year still crop up in Central Africa). The prevalence of this disease is due to persistent populations of a bacterium.

One reason bubonic plague was such a devastating killer in the Middle Ages was that no one understood its cause. People believed that the plague was caused by things like God's wrath, astrological events, or "corrupted air." They did not suspect that everyday animals and insects surrounding them were agents of disease. These creatures had never caused problems before, so why would they cause a problem now? Or so the thinking went. Steeped in teleology, the thinkers of the day believed that plague must be purposeful. It was easier to believe that God had sent the plague to punish sinners than to search for some "secondary cause" here on Earth. There were, however, no standards of sanitation in the fourteenth and fifteenth centuries; humans lived alongside rats and mice in nearly every "civilized" town and city throughout Europe, China, and the Middle East. Rat populations were huge, and the rodents lived on human table scraps, scurrying around homes with little to fear other than the occasional house pet. People had no suspicions that rodents carried disease. But indeed, in the fur of each rat were fleas, and in the guts of each flea were bacteria, later discovered and named *Yersinia pestis*.

The bacterium *Yersinia pestis* is benign to fleas.[87] It's an intestinal bacterium that prefers mammalian guts for its habitat. *Yersinia pestis* individuals essentially catch a ride in the guts of fleas that have bitten other infected mammals. They may spend a portion of their life cycle in the flea's gut, but the real action begins when the flea bites a new host—either human or rat—and regurgitates some of its gut contents, along with the bacteria, into the bloodstream of the animal or person being bitten. Rats and fleas are called vectors because they serve as agents of transportation for the bacteria as they spread to humans.

The original habitat (or "reservoir population") of *Yersinia pestis* is in rats and gerbils of Central Asia, and other wild rodents that live in semiarid climates.[88] The bacterium occasionally spreads outward from these population centers, usually causing only minor outbreaks when conditions are right. There are almost thirty different species of fleas that live on rats alone; they are ubiquitous and unavoidable. These fleas bite the rodents, ingest some pathogens in the process, then jump away to land on other mammals, bite them in turn—pathogenic individuals from their saliva—and spread the killing bacteria into the mammalian bloodstreams. All mammals that get bitten by fleas carrying *Yersinia pestis* provide a favorable habitat for the bacteria. *Yersinia pestis* settles into the lymph nodes, causing disabling pain and disfiguring swelling before death.

While bubonic plague persists today, it is now treatable. If left untreated, bubonic plague kills roughly 50 percent of those it infects. However, antibiotics can kill the bacteria in most cases, especially if the disease is diagnosed early. Still, 10 percent of infected people die even if antibiotics are administered. Pneumonic plague is even more deadly, causing death in virtually all who contract it unless it is treated by antibiotics within the first twenty-four hours of infection.[89] Even though medical treatment might save an infected person's life, hundreds of millions of *Yersinia pestis* individuals live on in other mammals.

We can't exterminate the bacteria. What we *can* do is minimize contact between rodents and humans with the goal of reducing fleabites. The only reason that we are relatively safe from bubonic plague is that modern cities have buffer zones between humans and rats. Municipal sanitation programs, sewers, pesticides, and exterminators all help keep plague-harboring rodents at bay. It's true that millions and millions of rats live in the tunnels and crevices of buildings in New York City, for example, but they are usually nowhere near the pantries or food supplies of humans. Whether these rats carry *Yersinia pestis* has not been well studied, but virtually all of them carry a heavy load of fleas, and all fleas can serve as a reservoir for the plague. Generally speaking, certain neighborhoods in cities such as Los Angeles seem to have higher risk for an outbreak—not because of the density of rats but because the fleas don't die off as quickly in that climate. The harsh winters in the urban neighborhoods of the eastern United States reduce the activities of insects and help to lower the risk of disease.

Today we obviously know much more about the plague-causing bacterium, *Yersinia pestis*, than did the citizens of the Middle Ages. We know about reservoir populations and the vectors of infection (fleas). In each blood meal, the flea picks up *Yersinia pestis* from the rodent that it bites, harbors the pathogen in its gut, and then regurgitates the bacteria when it bites another mammal. Eating a diseased rodent can also cause infections. The bacteria remain viable even in a newly killed animal.

Recent research suggests another interesting twist to the story. Scientists are now learning that the disease spread much faster than we previously thought. A new discovery suggests a human-to-human airborne transmission in addition to the slower process of animal and insect intermediaries to spread the bacterial population. *Yersinia pestis* doesn't enter the lungs until the infection has advanced and established itself in the human host. When it enters the lungs, bubonic plague becomes known as pneumonic plague, at which time it becomes infectious if the infected individual coughs. Today's antibiotics prevent the disease from "going pneumonic" because they control the infection quickly if people come down with bubonic plague. This keeps the bacterial population from exploding in number by restricting its transmission to fleabites. As stated earlier, if you control the reservoir population (fleas) you can reduce the human infection rate.

Even though we think of the plague as a problem of the Middle Ages, the disease is still viable. All the elements of the plague—*Yersinia pestis*, its vector, the common flea, and its reservoir population, gerbils and Old World mice and rats—are all very much alive in the world. Plague outbreaks are rare because scientists and public health officials remain vigilant, constantly monitoring disease reports and wildlife statistics. This is combined with the modern sanitation practices we consider normal today. If we lost any of these safeguards against rodent/human intermingling, another serious outbreak of plague might occur.

The reason that we remain susceptible is twofold: (1) Most humans lack the antibodies for plague. (2) The pathogen, *Yersinia pestis*, has a physiological adaptation that allows it to avoid phagocytosis.[90] These two factors conspire to make the plague a persistent threat to human health.

In order to understand how cell populations interact during infections, we have to understand how the immune system functions. Look at our Black Plague example: Those who suffered most from infection during the plagues in Europe were poor overworked laborers who lived in dirty conditions and in close communion with vermin. Immune systems don't work so well if their owners are poorly nourished, living in filth, overworked, and poorly rested. Therefore the progression of a disease like the plague, with a normally low incidence of infection, can become much more virulent when it shifts to human populations with compromised immune systems. Since health standards have improved worldwide there is less opportunity for *Yersinia pestis* to cause pneumonic plague. But as conditions of public health deteriorate, as they do in poverty-stricken parts of the world, such as Central Africa, and as humans there suffer more, immune systems will break down and provide a fertile habitat for multiplication of airborne transmission of microbes.

This is also happening with increases in tuberculosis. In certain parts of Africa, within the confines of single communities, the poor and weak succumb to tuberculosis more readily than the strong and well fed. This is attributable to the strain on weakened immune systems that comes from hardship and unhealthy conditions.[91] The immune system is an inheritance from ancient ancestors (as will be seen, certain substances of our disease-fighting arsenal come from animals that are only distantly related, such as sea urchins). Nonetheless, this ancient organ has been modified greatly throughout its history. The human immune system works so well and is so efficient at eliminating most pathogens that it's hard to find any failings in it. But it does fulfill the popular notion that it is an organ hell-bent on vanquishing an enemy population completely. It could better be seen, however, as a population of cells that requires education and occasional encounters with intruders in order to protect and serve the larger organism. I think the latter view is more realistic, even if it's a bit harder to wrap one's head around. But in order to explain the workings of the immune system, and to arrive at this more enlightened view, I shall proceed with the easier alternative, until the narrative device of warfare no longer serves us as a way to understand the immune system's function and development.

Think of what the most terrifying danger to humanity might look like. Forget genocidal maniacs, organized terrorists, or nuclear war. Even the worst human-made atrocity will leave survivors. But microbes are almost perfect villains. Theirs is an army of barely detectable living things surrounding us, wishing to live inside and on us, and parasitizing our crucial nutrients, cellular resources, and the genetic machinery of our own bodies in order to multiply and disperse themselves at the expense of our health and well-being. They seem downright "evil," in that they appear committed only to the continuation of their own species. I find nothing wrong with regarding them as my enemy. If the purpose of war is to annihilate evil from this world, then clearly my immune system is best understood as a fighting force on the side of good. The foot soldiers in the war of microbe against human are cells of the immune system, for the battle takes place on a field at the scale of nanometers and involves direct cell-to-cell contact.

Every animal has an immune system. Hence this epic battle of good vs. evil (health vs. infection) is going on constantly all over the world, in every habitat. Furthermore, since all organisms are related through common descent (the fundamental principle in evolutionary theory), we know that the battle has raged since our common ancestor roamed the primordial oceans sometime before 540 mya.[92]

The basic structure of the human immune system is composed of two parts, the "innate" immune system and the "adaptive" immune system. So-called primitive animals, such as sea urchins, clams, worms, and insects, have only an innate system, whereas the animals with "backbones" (all known as vertebrates) including fishes, amphibians, reptiles, mammals, and us have the additional "adaptive" component that works in tandem with the innate immune system. This adaptive immune system is not as old as the innate system. The common ancestor of all vertebrates probably was the evolutionary innovator of this vast army of specialized killing cells.

This struggle between the forces of good and evil has been going on for as long as animals have been around on the planet. And if the previous chapters have taught us anything, they have given us the reason why this is so: Bacteria and viruses have always been lurking and creeping into every possible habitat. These evil forces had a strong foothold on the planet even before the evolution of animals. Therefore, in order for the animals to evolve at all—and bring the forces of good to the biosphere—they had to overcome an inherent pervasive evil that existed for more than a billion years prior to their arrival. The armies of infection and plague ruled the planet when the biosphere was young, and early animals had to contend with this challenge before any other adaptation was possible. (You see here how easy it is when speaking of the immune system to lapse into almost biblical rhetoric!)

The most fundamental part of the immune system that is shared by humans, their vertebrate kin, and invertebrates such as horseshoe crabs, marine polychaete worms, starfish, and sea urchins is a protein known as complement.[93] Complement is a toxin produced in the liver and is secreted into the bloodstream whenever an invading microbe or foreign collection of cells enters our bodies. The point of infection, be it a puncture of the skin or a lesion inside the mouth, produces a series of chemical reactions; the resting cells of the liver are "called into action" to produce their special toxin, and it is released liberally into the bloodstream. This toxin eventually reaches the infectious population of microbes and envelops them.[94] In most cases the complement system destroys the pathogens by tearing holes in their cell membranes. This not only kills the invading microbe but also prevents it from reproducing and building up the army of infection.

But not all microbes go down so easily. Sometimes complement simply surrounds the cell membrane of an infectious microbe without actually destroying it. In such a case, when a complement molecule (or antibody, as we'll see below) comes into contact with a microbial pathogen, the pathogen is referred to as "opsonized." "Opsonin" is the general term for molecules in the blood serum that make pathogens susceptible to phagocytosis. The complement-encrusted pathogen floats aimlessly in the bloodstream, where it is then discovered by another branch of the innate immune system's military forces-for-good, the macrophage.

The macrophage is an ancient type of cell. Our bodies are full of them, especially in tissues that are likely to come into contact with pathogens, such as the mouth cavity, lungs, and the digestive tract. Also found just below the skin, in connective and lymphatic tissue, they form a loose accumulation, almost like a row of sentinel soldiers standing at arm's length of one another, surrounding the inside of the castle wall. If anything should pierce the battlement—in this case the skin or the lining of the intestines or lungs—then the macrophage soldiers quickly snap into action and destroy the invader.[95]

The macrophage—the word comes from Greek roots meaning "large" and "to eat"—kills bacteria and other invaders by simply engulfing them. The method of annihilation points to a primitive ancestry.

Even though we humans think of ourselves as the most highly advanced species on the planet, our innate immunity depends on cells that resemble the lowly amoeba.

One of the first things students are shown in biology class is what pond water looks like under the microscope. As a high school student I remember being amazed to see how many different kinds of living things were floating around in a small drop of water we collected from a nearby pond. I saw fast-moving paramecia dart across the field of view, those ciliated single-celled organisms that have photosynthetic pigments and rudimentary mouthparts. I saw some kind of colonial multicellular protozoan as well. Nothing was more interesting, however, than the largest cell on the slide. It moved by barely perceptible contortions, slowly morphing its outer membrane in one direction or the other, sending out extensions seemingly to make contact and "grab onto" much smaller foreign floating objects. This was the amoeba.

Amoebas make contact with foreign particles by changing their shape through a process of constriction and cytoplasmic fluid flow. To envision this, think of the amoeba as a tiny water balloon. If you squeeze and constrict a round balloon you can make it contort into almost infinite number of shapes. However, the amoeba's active phagocytic state looks more like the early stages of an animal balloon with a roundish body and a neck protruding from one side. This neck is called a pseudopod, and it extends in the direction of a piece of food.

Once the amoeba's pseudopod makes contact with a particle, such as a bacterium, it draws the particle closer to its "body" and begins the process of phagocytosis—eating the particle by surrounding it with its cell membrane. To envision this, imagine that you push a marble[96] into the side of the fluid-filled balloon until you nearly rupture the balloon membrane. As you push deeper into the side, what happens? The balloon "wants" to retain its shape as much as possible, so as you push deeper toward the center of the body, the balloon's membrane envelops not only your finger but also the marble, surrounding it to ensure that it cannot exit.

Now imagine this happening without a finger pushing the marble, and that the balloon itself is responsible for the contortions of its own outer membrane. This is essentially how phagocytosis works. The amoeba has a submicroscopic movable framework of molecules called the cytoskeleton that lies underneath the cell membrane. All eukaryotic cells use their cytoskeletons for cell division, but amoebas use this to make contortions far greater than those of other cells. Once the particle is drawn in and surrounded by membrane, it gets digested. It's almost as if the amoeba creates a stomach every time it engulfs a food particle or an enemy organism. After the engulfed contents are killed and digested, the reverse process takes place—exocytosis—and the waste product is excreted through the membrane back into the pond water whence it came.

The whole process can take an hour or more. I was an impatient youngster, so I didn't wait around to watch this entire process take place, but I remember my teacher showing me amoebas in different stages of their phagocytosis. And even though my teacher wasn't a professional taxonomist (she couldn't name the exact species I was observing), she made it clear to me that this organism was very primitive. I immediately understood the implication: Phagocytic cells have been around a very long time, longer than the dinosaurs, longer than the fishes, longer than the insects, longer than the plants.

It took me many years to discover that amoeba-like cells, whether primitive pond dwellers or immune-system sentinels, are all in fact evolutionarily related with respect to their mechanisms of phagocytosis.[97] All phagocytic cells carry a common genetic mechanism in their DNA that allows them to eat and sequester other pathogenic cells. This mechanism is so useful that it has been there all along, since the advent of the eukaryotic cell some 2.1 bya. This means that the foundations of our immune system—macrophage cells—carry the same genetic instructions as those little pond creatures I observed back in high school.

Of course the genetic instructions for this cellular mechanism are contained within our chromosomes and passed along from generation to generation on gametes (sperm or egg). Thus they have undergone countless, incalculable genetic changes since they appeared in our evolutionary ancestors. This means that a human macrophage is not identical to an amoeba from pond water. You couldn't, for instance, inject pond water into your veins and increase the population of macrophages in your immune system by forcing amoebas into a new environment.

The important implication from the evolutionary connection between macrophages and amoebas is that populations of cells we find today did not arise spontaneously to perform a particular service for humankind. Because of evolutionary descent, all organisms are related to one another, and this principle extends to the cellular level. There is an evolutionary progenitor for every kind of cell, just as there is for every species. Most of the cell types that make up the human body are ancient. Our cells have been performing the same kind of activities since long before our species came to exist. The way they come together to perform their synergistic function, however, is unique. The coexistence of these ancient cells is what makes us who we are.

When a new species arises, it inherits the organs of its ancestor, slightly modified, perhaps, but generally functioning nearly identically to that of its immediate evolutionary progenitor. The saber-tooth cat (*Smilodon sp.* from roughly twenty-five thousand years ago) had a retina that was very similar, if not identical, to those we find in house cats today. Likewise, the muscle cells that made up the quadriceps muscle on a carnivorous dinosaur (such as *Tyrannosaurus sp.* from 175 mya) would have been nearly identical to those we find today on fast-moving large reptiles such as the Komodo dragon. There is a fascinating evolutionary stubbornness to most cell populations; they simply don't change as quickly as species do. The organisms that harbor them come and go, but the cells themselves remain the same. It seems as if cell populations would rather mingle than evolve.

Most of vertebrate evolution is characterized by rearrangements of tissue and organ types and changes of skeletal structures as opposed to new organ "invention." Our evolutionary heritage is one of new functions for old systems rather than novel evolutionary "innovation."

Let's revisit lungs in more detail. Lungs went through a major shift in evolution about 300 mya when quadruped (four-legged) ancestors (called *labyrinthodon* amphibians) began to leave their shallow-water habitats to explore land. In order to live on land, first and foremost, animals have to breathe air. Originally amphibians were excellent at breathing water, like fishes; they did this by using gills and other highly vascularized tissue (like skin in amphibians, for instance) that can extract oxygen and offload CO_2 into water. Primitive lungs developed from swim bladders in lungfishes, the ancestors of amphibian quadrupeds and of all terrestrial vertebrates as well. Lungfish began "experimenting" with breathing air rather than water during the Devonian (419–359 mya).

They had no well-developed lung tissue, only highly vascularized swim bladders that performed the feat. Swim bladders arose in the lungfish ancestors to perform functions of buoyancy. Saclike, developing from an outpocketing of the esophagus, and located "below" the dorsal fin, these organs were pervaded by blood vessels that transported gases into the swim bladder to inflate it when buoyancy was needed.

It's always fun to speculate about the first lungfish ancestor that stumbled on the fortuitous activity of filling its swim bladder with air instead of gases from the bloodstream. At first the process must have been highly rudimentary. But some advantage must have been afforded. Perhaps it lived in a population that was restricted to stagnant water with low oxygen levels, so conventional gill breathing was difficult. In such a case any oxygen extracted from air would allow that lucky mutant individual who could breathe air to have more offspring and pass on the trait.

Seen in this light, the advent of lungs, then, was simply a fortuitous accident of evolution that allowed certain fish to make an excursion onto land. It was not a purposeful "invention" by anyone or anything in order to serve some ultimate purpose. When we use the word "innovation," we often assume this teleological notion. The emergence of lungs was an evolutionary rearrangement of an organ, rather than a designed innovation of new tissue by some crafty ancient organism. The cell populations that made up the swim bladder remained intact, but the individuals of those populations had different neighbors, and different means of communicating with their environment (gas exchange from the bloodstream or the air) in order to serve the larger organism. Blood vessels increasingly took on a new role to absorb oxygen and offload CO_2, so an old organ developed a new use. This is typical of evolution: Old organs and tissues come together in new ways to form new species while their individual cells retain their functions through time.

Back to the case of our immune system, then, where macrophages use the inherited "machinery" of their amoeba-like ancestors to accomplish their phagocytic function. The macrophage exists today, and has persisted through numerous eras in Earth history, performing its immunity function dutifully within the confines of hundreds of thousands of species. All animals with some sort of a circulation system have macrophages. Think of them as reservists performing their service for a purpose greater than themselves. Whenever a foreign substance is encountered or a pathogenic microbe breaks through the first line of defense (the skin or a mucous membrane), the macrophage is activated and begins its phagocytic contortions, first grabbing the particle, virus, or bacterium, and then enveloping it completely.

Once the macrophage has enveloped the foreign object it attacks it with lysotic chemicals. These work like digestive enzymes to dismantle the pathogen and neutralize its toxins. After the enemy is subdued and made harmless, it is either released as a waste product back into the circulating body fluid and excreted, or portions of it are retained on the surface of the macrophage.

Unlike their primitive ancestors (the amoebas), macrophages are not actively "hunting" for food. Instead they become activated when an invader pierces the protective veil of the skin or mucous membrane. They lie in wait for months at a time, essentially resting until they are presented with an invader by associated immunity cells (we will look at these later) or directly perturbed by a pathogen.

But pathogens, like all military enemies, have evolved mechanisms to avoid destruction. Like covert operators, some, such as the *Yersinia pestis* bacterium mentioned earlier, break into the bloodstream through mucous membranes and are not "recognized" as pathogenic. Like disguised infiltrators, they circulate and mingle with macrophages but have special molecules that interfere with the macrophage's cytoskeleton, rendering it unable to contort and engulf foreign objects.[98] They avoid phagocytosis altogether and use the host animal's blood and lymph as a habitat for reproduction. The macrophages don't recognize *Yersinia pestis* as a pathogen. Instead, new pathogenic clones of *Yersinia* are formed, and the virulent offspring emerge and take up residence in the lymph nodes or lungs. This leads to a massive buildup of troop numbers in the infecting microbe population, and debilitating suffering in the host.

Despite the rare occasions of pathogen elusiveness as described above, the general efficiency of macrophages is remarkable. The macrophages scarf up pathogens and also consume anything that is not crucial to the proper functioning of the host organism itself. For instance, most of our cells have a limited life span; one million cells die every minute in the human body (interestingly, cells in the cerebral cortex are among the few that live for the duration of the organism's life[99]). Cells that die inside our body are cleaned up by macrophages. These include cells damaged by viral infection, as well as those that are killed by physical abrasions in the intestines or deep cuts into the epidermis. These dead cells must be removed, and the macrophage, as part of its normal resting activity, performs these sanitation duties.

The macrophage has other jobs in addition to general cleanup of dead cells. In the aggressive counterattack against increased pathogenic activity, macrophages release deadly chemicals that kill nearly all invaders. Tumor cells, for instance, are destroyed by one of these chemicals, called tumor necrosis factor (TNF). The macrophages also secrete hydrogen peroxide to destroy multicellular parasites. Once the macrophage is activated it essentially sounds an alarm that rallies other components of the immune system into battle. Furthermore, macrophages have another trick up their sleeve that is revealed when they are in their activated state.

Macrophages become "antigen-presenting" cells when they are activated. In this phase of their activation, macrophages ingest a pathogen and dismantle it protein by protein. But instead of simply secreting the remnants of the infectious microbe, the macrophage selects the most unique portions of the pathogen and moves them to the cell surface, where they can be "presented" to other soldiers of the immune system. In the ancient world, armies would display the heads of their enemies on spears. This served a purpose; it reminded the troops of whom to kill in battle. The macrophage is essentially doing the same thing: educating its fellow warriors on the new threat to the organism.

Our innate immune system is derived from a remote pond-dwelling ancestor. For hundreds of millions of years, rapidly evolving microbe populations have been coming up with ways to avoid being engulfed by these amoeba-like cells. As remarkable as our macrophages are, in the guarding of our health, they need support from other cell populations. Macrophage ancestry is so ancient that there have been countless numbers of pathogenic "inventions" to outwit them. Some, such as *Yersinia pestis,* are living examples of this, as illustrated above.

Neutrophils are the other kind of "professional phagocyte" that make up our innate immune defense. Like macrophages, these cells come from stem-cell progenitors in our bone marrow. Whereas macrophages generally "lie in wait" in the tissues, neutrophils are found actively circulating in the blood stream and lymph. In fact 70 percent of the white blood cells you find when a doctor takes a sample from your vein for lab tests are neutrophils. They are more numerous than macrophages, but also are much shorter-lived. Neutrophils live only about five days (macrophages can live for months), undergoing a planned death after a brief but devastating life.

Neutrophils are voracious and deadly. There is no subtlety to these "soldiers." They simply release toxic chemicals (such as TNF), eat everything in their path, and then die. Neutrophils have no antigen-presenting function or chemical "call-to-arms" signals. Because of this nonspecificity, neutrophils are key components in one of the most familiar types of immune reaction, the acute immune reaction, or "inflammation."

When we injure ourselves we open up a hole in our first line of defense against the evils of ubiquitous invading microbes. Bacteria and viruses are everywhere, inhabiting the surfaces we come in contact with and populating them by the billions. There are literally billions of opportunistic infecting agents surrounding you at any given moment, lying in wait until that unfortunate moment that you cut your finger or toe, or face-plant your lips in their vicinity. These injuries introduce these pathogens into a new habitat, your bloodstream, into which they can migrate and proliferate along the capillaries and connective tissues of your skin. Even something as tiny as a pinprick can introduce hundreds of kinds of pathogens into your body.

Luckily our cellular defenses are prepared for such instances. As soon as the injury takes place, circulating neutrophils leak out onto the bloodied surface and begin their slaughter of every possible invader. Even if blood is not spilled, as in the case of bruising or abrasion or joint and muscle fatigue, a noticeable redness and pain and burning sensation is associated with most types of familiar injuries. These three factors—redness, heat, and swelling—are called inflammation. Inflammation is a common cause for a visit to the doctor. Anything with an "-itis" for a suffix—such as bronchitis, tonsillitis, arthritis, dermatitis, and so on—describes a condition where neutrophils have been called into action.

Neutrophils swarm in huge numbers at the injury during an acute immune reaction. It takes them about thirty minutes from the moment of injury to get there; they leave the bloodstream, prying their way out of arterioles (capillaries) near the damaged tissue. If a puncture wound or scrape is deep enough, the neutrophils can also leak out of damaged blood vessels. It's neutrophils versus bacteria, and they are getting ready for an epic showdown; our "good" neutrophils are triggered to act by chemical cues on opsonized bacteria.

The more complement that binds to bacteria, the more neutrophils respond and the more intensive the phagocytosis. The neutrophils show no mercy, attacking all free cells in the area that carry the label from complement. All this action can have negative consequences. This intense phagocytic activity also causes exocytosis—the moving of substances out of the neutrophil—which in turn causes destructive chemicals to flood all cells in the area. Neutrophils can pour forth dangerous amounts of oxidizing agents, such as hydrogen peroxide, that not only kill bacteria and cells infected by viruses but also can make healthy tissue red, sore, and functionless.

The body rids itself of excess and dead neutrophils during inflammation. When a wound leaks it is partly because of this process; in fact most of the makeup of pus is dead neutrophils.[100] There is another biological mechanism, however, that limits the destructiveness of neutrophils. As mentioned above, neutrophils live for only about five days. Unlike macrophages, which take up residence in the connective tissue and live for months, waiting for the enemy like entrenched sentinels, the neutrophils act quickly and then commit suicide. They undergo "programmed cell death" soon after their phagocytic and cytolytic activity is at its maximum. These suicidal neutrophils get flushed from or weep from wounds, or circulate and are removed by digestive processes, and the tissue repairs itself after their departure.

So far I've been satisfied using the war metaphor to explain the immune system. All this activity is easily framed in terms of conventional human warfare. Microbes and pathogens, like the culprits of former president George W. Bush's "war on terror," are ubiquitous on the planet, inhabit every possible surface from soil to needle points, and are also ingested with every breath we take because they are floating in air. It's easy to paint them as "the force of evil in the world." In this light our innate immune system functions purposively to rid our bodies of evil in order to maintain "good," which in this case is a healthy body.

We have various weapons to combat evil. Some are chemical "weapons," in the form of complement proteins that float about in the bloodstream ready to latch onto invaders and make them "visible" and easily detectable for dismemberment by our innate immune system's foot soldiers. Our "troops" of soldiers have already been deployed in advance of any such evil invasions: Macrophages live in our muscles and connective tissues, and neutrophils circulate in the bloodstream. Both kinds of troops can kill evil invaders with "hand-to-hand" combat (phagocytosis—the engulfing of pathogens) or with their own form of deadly "hand-grenade" deployment (the releasing of toxic chemicals, such as hydrogen peroxide in neutrophils).

With neutrophils, however, we come to a flaw in the war metaphor. Every time we bump or bruise our skin, each time we rub our eyes because of itchy irritations, we are activating an army of suicide bombers. Such fighters don't fit the conventional concept of victors in the time-honored narrative of conventional war. In fact, history has yet to reveal a lasting empire made stronger by a standing army of suicidal troops.

What happens when we rub our eyes during allergy season? Do we soothe our irritation? No! We just make it worse. Our eyes get redder and they start to burn. The sensitive eye tissue responds to our careless handling by swelling and oozing. As we continue rubbing we inadvertently recruit neutrophils to spill their toxins on the affected region, injuring healthy tissue with their highly acidic secretions, in addition to killing any possible pathogens in the region.

Neutrophils are neither "with us or against us," as Bush famously said. Rather they exist, fulfilling an automated function without the ability to consider whether doing so is the best course of action. This takes us back to an earlier idea—that our bodies are more a loose coalition of distinct organs and cells than a singular "self." If one organ (or one subpopulation of cells within this organ) is malfunctioning, it can lead to the demise of the entire organism.

So before we get too carried away with our war metaphor, we have to remember the essential facts—and nonteleological nature—of biology. None of these systems were created specially for us. There is no "grand designer" who orchestrates infections, plagues, or pandemics or engineered our defenses to them. All these mechanisms that we attribute to a battle between good and evil are in actuality biological traits that we have inherited from preexisting populations. Therefore the interactions we are witnessing (infection, inflammation, phagocytosis) are based on previously established conditions of coexistence, and we should not expect to find any sort of unique perfection in our immune system. After all, these systems are not at some end point of evolution; they are still evolving.

Rather we should expect to find ancient cellular systems from distant ancestors that have come together to work synergistically.

We know that human macrophages inherited their cellular "machinery" (cytoskeleton) from amoebas. But there are numerous more recently derived organisms, such as sea urchins and slimy intertidal burrowing fishlike animals (called lancelets; genus *Amphioxus*), that have well-developed innate immune systems that are very similar to our own. This is strong evidence that the types of cells involved in immunity—like the organisms that harbor them—are linked by evolutionary descent, since the organisms that carry them have been shown to share common ancestors. In fact all macrophages undergo a developmental process inside our bodies that mimics the evolutionary changes of our ancestors in deep evolutionary time. Roughly 4 percent of the white blood cells circulating in our blood are called monocytes. These cells are attracted to sites of infection or injury, leave the bloodstream, and enter the damaged tissue, where they "mature" into macrophages. Hence monocytes represent a stage of evolutionary development that is more primitive than the macrophage.[101] These cells represent the "living link" with our primitive ancestors, and they prove that our own defense system is a modification of those from past immune systems.

At some point in vertebrate evolution, probably before 500 mya[102], a new layer of protection emerged to function in concert with innate immunity. Humans, like all our vertebrate ancestors, have a second type of immune defense that works in unison with the innate immune system: the adaptive immune system.

The adaptive immune system is a marvel of biological diversity. If we apply a military analogy, is the most sophisticated type of defense system ever developed. It is like a Central Intelligence Agency that knows everything about possible threats to our security and works with seamless efficiency alongside the Pentagon, deploying troops as needed to quell each type of specific threat. In fact, it's another suite of white blood cells that have greater specificity than the macrophages or neutrophils of the innate immune system.

True adaptive immunity in humans is a function of two types of cells: B-cells (aka B-lymphocytes, which mature in the bone marrow) and T-cells (aka T-lymphocytes, which mature in the thymus). Part CIA agent and part chemical-weapons manufacturer, these cells target specific microbial pathogens and deploy in a syncopated manner to assist the innate-immune-system troops in finishing off the enemy. Their main weapon is called the antibody.

Antibodies are molecules created by lymphocytes, more specifically, B-cells and T-cells. Generally speaking, we can think of B-cells as antibody factories that produce huge amounts of antibodies when they are "activated." In some instances they can pump out two thousand antibody molecules per second.

Like neutrophils spilling cytotoxic chemicals all over the place, B-cells spill millions of deadly antibody molecules into the bloodstream. B-cells can be found freely circulating, or lingering in lymph nodes, or embedded in connective tissue (where they are called plasma cells). Unlike neutrophils, however, the antibodies created by lymphocytes (B-cells and T-cells) have specific targets. Whereas the toxic substances of the neutrophil are damaging to all cells in the region, antibodies affect only specific pathogens.

T-cells also produce antibodies, but rather than being mass-producers of molecules like an activated B-cell that spills its products liberally, the T-cell retains its antibody on the cell surface. T-cells work in partnership with other proteins and in some instances with macrophages of the innate system that have ingested bacteria. These proteins and macrophages change in response to the presence of pathogens and perform the important role of "antigen-presenting." Antigen-presenting cells bring pathogens to T-cells. When the appropriate T-cells recognize the prisoner, they make cell-to-cell contact and destroy the pathogen by tearing holes in its cell membrane.[103] If the prisoner is a bacterium, the T-cells have a direct impact on the invading army by killing its soldiers outright. But sometimes the prisoner is one of the host's cells that has been infected by a virus.

Remember that viruses live within the nuclei of formerly healthy cells. During viral infections T-cells destroy host tissue. That is to say, our own body cells, when infected by a virus, are destroyed by our own population of T-cells. It is a case of self destroying self, but brought about by a foreign virus.

The specificity of lymphocytes is their most amazing property. It's almost beyond comprehension: For every possible microbe and pathogen, there is a corresponding antibody produced by our lymphocytes. It's important to note that antigens can be chemicals encountered in the environment, or proteins on the surface of unicellular organisms or viruses. Scientists believe that roughly 100 million different antigens exist in the world. This is a huge number, and of course no single organism ever comes into contact with that many in its lifetime. But in order to do its job, the adaptive immune system has to provide immunity in many unforeseen circumstances in case new antigens are encountered. That means that 100 million varieties of antibodies can be produced by the adaptive immune system. The B-cells have a unique way of doing this: It's called modular assembly.

To envision this, consider the proteins created by the human body. A protein is basically a chain of amino acids strung together. We have roughly twenty thousand different proteins that are produced by our cells during our development—for example, hair, skeleton, blood, hormones, pigmentation, muscle, and so on. Our cells use modular assembly of only twenty amino acids to make all these varieties. That is to say, the great variety of proteins is not due to changing the structure of amino acids, or due to inventing new ones, but by variations in the way those twenty amino acids are sequenced in the chain. Modular assembly creates tremendous diversity by reorganizing a much smaller subset of units.

B-cells have the capacity to create very highly diverse antibodies because they carry chains of subunits—sequences of DNA—that can be arranged and reordered by modular assembly. There are four types of modules that are strung together, but each one of these modules has between six and twenty-five varieties. Roughly ten million different molecules can be produced from these four modules just by reordering them. Adding and subtracting DNA bases at the junction between modules adds even more diversity. This junctional variation coupled with the modular assembly of the other subunits is enough to create more than 100 million different varieties of antibodies. Susumu Tonegawa received a 1987 Nobel Prize for this discovery. According to immunologists, this number of antibodies sufficiently protects us from all known or conceivable antigens on the planet.[104] This is accomplished through the fundamental processes of population increase and specialized genetic recombination, as progenitor lymphocytes develop and increase their numbers.

It is crucial to note here that there are two sets of vessels in the body. The blood vessels are the most familiar because, when we cut ourselves we see red blood coming from them. This is caused by erythrocytes, also called red blood cells, that circulate constantly throughout the blood vessel network. There is, however, a completely separate network of vessels that carries clear fluid, so you'd hardly know if you ever cut one because you likely would focus on the red stuff coming out of the wound. Nonetheless this vascular network, called the lymphatic system, runs through your entire body and carries lymphocytes between the hundreds of lymph nodes, the thymus, and the spleen. Intimately associated with blood circulation, the lymphatic system empties its contents (lymph fluid from tissues, and lymphocytes) into the blood vascular system at portals near the base of the neck.

Lymphocytes mature in the thymus and the bone marrow and then migrate to lymph nodes of the lymphatic system. You are probably familiar with lymph nodes because whenever you are fighting an infection, such as a sore throat, your doctor feels your neck just under your chin, looking for swollen lumps. The swelling is caused by populations of T-cells and B-cells proliferating due to the presence of an infection. Both types of cells can exit the lymph nodes and circulate in the bloodstream. Cycling between the lymphatic circulation and the bloodstream may take weeks or months for each individual cell. All the while lymphocytes are dividing and accumulating antibody variations that will match particular antigens. (When B and T cells divide they undergo a rearrangement of their genes. The instructions for making antibodies is one portion of the genome, and it is particularly active in this process of genetic recombination.)[105]

All this genetic variation in the antibody-producing region of lymphocytes results in a population of progeny that is highly variable. So variable, in fact, that more antibodies are produced than the number of antigens the organism will likely encounter in its lifetime. The excess antibodies are like an invented weapon that will never be deployed. As mentioned above, it is estimated that we carry 100 million different kinds of antibodies; each one is specific to a single pathogenic organism. This seems an odd arrangement for an army, but such potential serves us vertebrates well. It allows us to inhabit a fantastic variety of habitats, each with its own endemic pathogens.

The specific suite of antibodies that are retained and amplified is a product of environmental selection—natural selection, actually. It is the key to the adaptive immune system. At any one time, for instance, there are only roughly ten thousand T-cells for any specific antigen circulating in the entire body. This is an extremely diffuse population of individual cells (consider that ten thousand cells makes up only a tiny portion of a microliter of liquid, visible only under the microscope). With such a diffuse distribution of cells carrying a specific antibody, it is unlikely that one of them will get "activated" by coming into contact with a pathogen that has the appropriate antigen specificity. If we get attacked by a sizable population of microbes, in a region of the body that is devoid of appropriate T-cells, the infection could be devastating. Furthermore, the antigen has to be presented to the T-cell in a nearby lymph node by an appropriate "antigen-presenting cell" (APC) that has come into contact with the invading microbe.

The lymph nodes, however are distributed throughout the body at places where circulating APCs introduce T-cells and B-cells to the foreign invaders. After the introduction, the T-cells and B-cells become "activated." That is to say, a cascade of chemical cues snaps them into action. Remember that only one subpopulation of lymphocytes—those bearing the proper antibody—becomes activated.[106] After the successful match is made with only a tiny subset of the lymphocyte population, a single variety out of hundreds of thousands of possible antibodies, a process called "selective cloning" begins.

Selective cloning is the key to adaptive immunity. After the particular antigen/antibody complex is formed, cell division takes place. In the case of clonal selection, every daughter cell is identical to its parent. Each time the population doubles, therefore, there is a doubling of cells that are committed to making identical antibodies that specifically match the chemicals on the surface of the pathogen. These antibodies are like heat-seeking missiles. There is no escape for the pathogen. They all contain a specific antigen "label" on their cell surface that says to the antibody, "Here's the bull's-eye."

As the pathogen population diminishes, there are fewer bull's-eyes, and most of the B-cells and T-cells specific to the invader die off as well after they deliver their deathblows. But not all of them die off. Some are kept circulating as "immunological memory" cells. For instance, T-cells, with specific antibodies, have been found to live for ten years or more. Clonal selection of these immunological memory cells ensures that if ever a particular pathogen is encountered in the future, its antigens will be "recognized" by these mature lymphocytes that can activate quickly to mount a specific adaptive immune response.

The war metaphor runs into further problems here. It turns out, for instance, that what really brings about the immune reaction is not some kind of strategic objective of our immune system to annihilate the enemy for resources or territory. It is purely a defensive chemical reaction. And further, this deployment of troops is unique and unlike anything witnessed in the history of military action.

With clonal selection we see, finally, that the war metaphor has lost its utility. No army in history has ever functioned with such elaborate, cumbersome artillery as the human adaptive immune system. For the millions of possible "enemies" (pathogens) in the world, the vertebrate immune system produces a specific "solution" for each one of them. Like undetonated bombs, lymphocytes with particular antigen specificities circulate in our blood and lymph systems. Each one of them carries an antibody that in all likelihood will never meet its "fated pair"—the exact variety of pathogen that matches its antibody. It's like an army with literally millions of specialized troops, most of whom will never see combat. If one of these specialists ever gets called to arms, he is expected to create an entire battalion as clones of himself, and then go out and finish off the enemy. There are no appropriate war stories to characterize such an odd arrangement (though it might make an excellent premise for science fiction).

Wouldn't it be better, for instance, if instead of carrying around all those "unused" B-cells—the ones specific to pathogens that we don't encounter—we could transform a portion of our adaptive system into an innate system that is nonselective? In this way we could keep all common pathogens at bay in whatever environment we chose to inhabit (remember that macrophages and neutrophils don't discriminate, they just scarf everything that is near them). It turns out that B-cells and macrophages are related evolutionarily, and B-cell progenitors can be induced in the laboratory to develop into macrophages. This means that cells once thought to have no phagocytic properties (B-cells) can quite naturally become amoeba-like phagocytes based on the environmental milieu. In fact, phagocytic B-cells are known to occur in some fish species, whereas human B-cells cannot perform such functions.[107]

Remember that the pathogenic environment "activates" the immune system. Without an appropriate pathogen stimulus, no adaptive immune reaction or inflammation results. Furthermore, it is the developmental environment that can determine whether a progenitor cell becomes phagocytic or develops into a lymphocyte. The "war" that we depict in our explanations of immunity and disease helps us to understand infection in a single individual, but it doesn't take into account the fact that the so-called enemy—in the pathogenic species—lives on in the host species in other individuals and perhaps benignly in host reservoirs. The most unfortunate fallout of this war narrative is that it leaves us with an unrealistic expectation that we can conquer disease through extinction of microbes rather than through population management of pathogens and immune systems. We need the pathogens to "educate" our adaptive immune systems. Hence, disease control should be seen as a commitment to stewardship. Instead it's billed as one of the wars, among an endless list of others, that can be won through vanquishing a perceived enemy absolutely.

It's crucial to remember that the immune system evolved by populations interacting with one another and with their environment. It is through the understanding of selective cloning and the modular generation of genetic diversity (from specialized recombination or directed mutation) that we can hope to manipulate our immune-cell populations in order to control pathogens. We cannot hope to eliminate all pathogens: If we did so we would have no way to "educate" our immune systems. But we can try to manipulate populations so that the effects of pathogens aren't so devastating. Immune-system education is, in fact, the goal of vaccinations—to train immune systems by introducing them, in a controlled way, to the very pathogens that could cause bodily harm or death. Without exposure to disease-causing microbes, an immune system is not "educated," and it can easily be overwhelmed by infection of a new pathogen.

What exactly is a vaccine? It's essentially the injecting of an antigen into our bloodstream in order to bring about an adaptive immune response. Antigens are contained on the cell surfaces of pathogens (or on the outer "envelope" membrane of a virus particle). We can "peel" a pathogen with high-speed blenders filled with bacterial or viral "soup" made of isolated cultures created in the laboratory. By injecting the shreds of these broken pathogens' membranes into our bloodstream, we are introducing an antigen into our system, mimicking a "natural" infection and "calling forth" the appropriate B-cell variety that carries the matching antibody.

Since the genetic material of the pathogen is not injected in a vaccine, there is no way for the harmful organism to reproduce. Only its antigen is administered—a potent protein that brings about our immune reaction, but as a separate entity it cannot lead to population increase. Our adaptive immune system jumps to attention upon discovery of this vaccine. One of the thousands of individual lymphocytes carrying the appropriate antibody makes first contact with the circulating antigen, which brings about a cascade of reactions that leads eventually to selective cloning.

But since there is no population buildup of the pathogen's population, the cloning of the lymphocyte only goes so far. A relatively small immune reaction results. The actual number of lymphocytes produced is somewhat larger than before, but not so large that it interferes with other varieties of B-cells or T-cells. With this limited population size, however, the adaptive immune system is "primed" in the event that a true infection should occur with that particular pathogen. Immunity from an infection results when our adaptive immune system has been exposed to an antigen and has undergone a slight population buildup of lymphocytes that are matched to that particular microbe's antigen.

Vaccinations are the application of evolutionary principles in action. If we can control the contact made between pathogen and lymphocyte populations, we can go a long way toward eliminating disease.[108] It doesn't require total annihilation but rather a control on population dynamics. Vaccines are the way we use selective cloning to keep a pathogenic population in a state of benign coexistence. The process is based on evolution, as pointed out by Nobel laureate Susumu Tonegawa: "Genes can mutate and recombine. These dynamic characteristics of genetic material are essential elements of evolution. Do they also play an important role during the development of a single multicellular organism? Our results strongly suggest that this is the case for the immune system."[109]

Instead of total destruction, which evolutionary history has shown is impossible (pathogens have been around since the beginning), vaccines demonstrate the more reasonable approach: Control the degree to which populations coexist, and you can preserve the necessary diversity of life. We don't tear down old cathedrals, we repurpose them. Our organs aren't created de novo for each new species; they have been inherited and reworked to perform new functions, even if sometimes less efficient than what an engineer might design. It's the persistence of coexisting diverse populations, cellular and organismic, that is a recurring phenomenon in the pageant of life. This is the message we need to use to educate people about their world, not the tired, outdated metaphor of victors and vanquished in some imaginary "war of nature."

The cells of the human immune system respond to the presence of alien pathogenic populations. The chemical reactions that play the key roles in the immune response are refined leftovers of ancient population wars that have been going on since the earliest days of the biosphere. With vaccination, we've come a long way toward demonstrating that stewardship of population interactions—in this case pathogens and their environment, the human immune system—can result in benign coexistence. How wonderful it would be if we could, on a global level, make the decision simply to control and occasionally limit population interactions rather than try to exterminate everything we consider as "evil."

7

WAR IS UNWINNABLE

America's most infamous war is the one we call "Civil." After its brutal conclusion, Gen. William Tecumseh Sherman told the citizens of the united Republic, in a speech to school kids, that war is all hell. The chronology of American history is usually depicted as a sequence of these hellacious episodes. Just as the history of the biosphere is revealed by the fossil record—communities of intermingling species interrupted by periodic mass extinctions—we trace our cultural inheritance by narratives of combat that interrupt periods of "peace."

In this chapter I intend to show that this history of human warfare has some of the same characteristics as the "warfare" between species that we see in nature. That is to say, war is an inevitable feature of all populations. If we view wars as the culmination of incompatible uses or acquisition of resources, it's easy to understand their inevitability in the course of human events. Just as populations in an ecosystem are thrust together in competition for limited resources (food, nesting sites, home ranges, and so on), human populations eventually intrude on each other as their numbers increase. Nations experience destabilization when this happens to their people, often resulting in destabilization of populations that were once in equilibrium.

None of us were actually present during these conflicts of old, yet we still align ourselves with one side or another. We know who the "good guys" were, at least according to our own historical circumstances. We can make a list of the "bad guys'" deplorable traits.

Conventional warfare tends to mimic some trends seen in wild populations. For instance, many human conflicts can be reduced to wars over resource utilization (think of oil in the Middle East). But human wars differ from population wars in nature because they are predicated as much on ideology as on natural resources (even though wars over oil have raged for decades, the political justification for entering the conflicts has centered on democracy, communism, or some religious doctrine). Ideas can spread like an infectious disease. Although this may sound like a poetic lyric, in fact thoughts are transmitted in culture just as biological traits are passed from one generation to the next. These phenomena simply use different mechanisms of replication and transmission. In the case of organismic traits, genes are the units of replication, and they are passed from parent to offspring. In the case of ideas or ideologies, symbols (words, gestures, rituals, and the like) are the mechanism of replication, and they can be passed from parent to offspring or from any two members of the same culture. Even though it can spread out of control, ideology is manageable to some degree, and therefore we have the potential to alleviate much human suffering.

Ideas spread quickly in our modern industrialized society: new ways of advertising, new products, and new behaviors, some useful, and some not. Our culture—like most populations'—is constantly changing. I like to believe that I can choose to ignore these changes and remain unaffected by them, but if they are too persistent (for instance, the use of the word "dude" or the belief that a moderate candidate is better than my favorite polarized one), I find myself "going with the flow" of things. We live in a culture of material upgrades that favor novelty over necessity. Most of us are aware on some level that these improvements exist simply to get us to spend more money rather than dramatically improve our lives. However, it's hard to resist the perceived excitement that comes with these purchases. Some people have to upgrade their perfectly functional smartphone as soon as a new model comes out. My weakness is tractors: I've owned five different ones in the last four years. My friends laugh when I explain how they each serve distinct functions, but I find it easy to justify the variety to myself.

I try to avoid the majority of movies, gossip magazines, and TV shows that dominate our culture because of their sheer, overwhelming banality. But the truth is that I sometimes get sucked into watching or reading them too; Justin Bieber, the Kardashians, Gordon Ramsay, the Bachelor, or the Bachelorette—it depresses me that I know who these people are. We are lured into reading about or watching them, not because they have anything at all to teach us about living a healthy or logical life, but simply because they have become part of our culture.

Regardless of the rational or intellectual efforts to resist them, fads, trends, and pop culture seem almost impervious to reason, and they quickly "go viral" and infect our daily lives. Some of them are quickly forgotten, like a flare-up of an infectious disease. But others persist longer and can become ingrained in the teaching and routines of our society. Think of the ways conversations have been altered, for instance, by the ubiquitous use of cell phones and tablets. Sure, they are good for communicating at a distance. But my daughter and her friends use them to communicate with each other even when they are in the same room! Often they are talking to one another and at the same time engaged in chatting to a third party via text on their phones. They don't even look at one another during conversations. Likewise, when I lecture, many students stare at their laptop screens, never making eye contact with me. I know that they are typing notes on the subject material, but I also know they are simultaneously browsing other things online and probably, like my daughter and her friends, engaged in multiple conversations. In the recent past such behavior might have come across as rude—even in the 1980s and 1990s, all eyes were on the lecturer—but this has become standard behavior in our modern, electronically connected society. It's all justified by the ideal of a connected social network and the ideology that computers are the best tools for a modern education and thus for the improvement of young people's lives.

Although it's debatable that computers have helped all citizens get smarter, the example serves to show that new societal norms can arise quickly. Once they do so, it becomes increasingly difficult to debate their appropriateness. These behaviors become tacitly accepted as more and more people engage in them. It's this unquestioned, ritualistic, habitual behavior that further validates the ideology.

We don't often think of war in terms of habitual or unquestioned behavior. Wars are supposedly carefully constructed political affairs, passionately deliberated, tightly constrained, and purposeful. Can we really see human wars in this light? Or are they better viewed as ideology gone awry?

The intellectual justification for war seems almost as important in human affairs as does the need for resources. This might be due to the anthropological tenet, mentioned earlier, that humans can be considered half biology and half culture. War is founded on principles of ideology perhaps because of our deep-seated need to act justifiably, in other words to behave in accordance with our culture. But in this chapter I'd like to address the possibility that we might also view warfare in a different light, as a by-product of the deeper biological imperative—population growth.

The geographical spread of people and their ways of life ultimately puts strains on basic environmental needs. Politicians have historically ignored this obvious reality. Perhaps this is because they are "political animals," and their focus on ideology fans the flame of their public support. But perhaps it is more important to acknowledge that scarce resources often spark the initial conflagration.

The default intellectual position taken by most people is that resources are scarce, and competition determines who gets what. This is based on an ideology informed by Darwinism that originated during the nineteenth century, as we will see in chapter 8. Today it has become tacit: "Without [the] struggle for existence, the agent of the selective process in the state of nature would vanish."[110] The idea here is that selection removes the unfit. But the facts of history trump the ideology. No empire or force for "good" has ever successfully eliminated a population of "evildoers." The populations we claim to have vanquished are still with us today and contributing to our society in ways that are usually unacknowledged. Perhaps the real "hell" of war is that you can never really win one.

What we usually think of as "vanquished" populations are really just groups of people much reduced in number and influence. But it should be remembered that populations, once established, have a strong tendency to persist. From infectious bacteria to predatory mammals, ancient enemies to modern crime organizations, populations tend to stick around, exert a continuing influence on our day-to-day existence, and lie in wait for circumstances to change. Once populations are established they are very hard to eradicate; they assimilate with more populous groups and new equilibria become the norm. This is true on a microscopic scale, as we saw in the previous chapters, and also on a macroscopic human scale as seen, for instance, in the complex historical accounts of regional "first contacts" between Native Americans and Europeans.

If we want to consider the human costs of population wars, then there is no better place to look than the turbulent and violent history of New York State, specifically the upstate region around the Finger Lakes.

In the centuries after the first contact between Europeans and Indians, settlers streamed into the seemingly underpopulated regions of New York, valleys now known by the modern names Hudson, Delaware, Mohawk, Susquehanna, and the central part of the state—the Finger Lakes region. The area around modern-day Buffalo was sometimes referred to as the Niagara Frontier. Early Europeans saw plenty of potential in these places: lots of seemingly underutilized land, timber to build with, and an abundance of game. European settlers observed what they considered an alien population, and judged them by their own standards of civilization or success—while the native people, the Iroquois, similarly judged the Europeans to be the strange aliens. The Europeans generally failed to see that the Iroquois actually had a well-established civilization that operated under clear social strictures. They were oblivious to the fact that the Iroquois carefully monitored and utilized the wilderness.

Many male Iroquois despised farming, seeing it as woman's work; instead they hunted and foraged in the woods. They were careful to hunt within reason, leaving enough deer to maintain the herd population. They were, in fact, stewards of the lands that settlers saw as an unmanaged wilderness. Their preferred way of life supported fewer people than intensive agriculture would allow, but they accepted this tradeoff in order to maintain a way of life that they preferred. The Indian nations didn't consider themselves the owners of the land, but each tribe had agreed-upon home ranges. This philosophy was so different from how the settlers viewed the land that it was nearly impossible for either population (Indian or white) to understand the other's perspective. The settlers saw the Iroquois as lazy savages who were too indolent to make the most of their verdant land. When the Iroquois eventually made land deals with the settlers, they assumed that the Europeans understood that their new property came with an obligation to assist the Iroquois with food (and eventually rum) in perpetuity, and (in most cases) that the land was being leased rather than sold. This fundamental misunderstanding caused perpetual social and political problems.

At first the Iroquois Confederacy benefited from "middleman" status between their Indian neighbors in Canada and the Europeans—France to the north, Britain to the south. Their geographical location meant that they were literally in the middle of two distinct imperial conflicts that bled over to the New World from Europe. As the British and French Empires grew hungrier for the territory and natural resources of North America, the Iroquois found themselves in a geopolitical dilemma. The French concentrated their efforts along the northern bank of the St. Lawrence River, starting in the seventeenth century, establishing trading posts that became towns, such as Montreal, Tadoussac, Trois-Rivières, and Quebec. The English acquired Fort Orange, New York, (later called Albany), an important trading post and later diplomatic center on the Hudson River, from the Dutch in the seventeenth century.

The Iroquois were then one group of five nations that made up only a portion of the large mosaic of ethnologically distinct native groups throughout the Great Lakes and northeastern parts of the New World.[111] Though numbering fewer than twelve thousand or so individuals around the time of first contact with Europeans in the 1530s, the Iroquois somehow were able to control the region east of Lake Michigan by decimating or displacing their surrounding tribes and wresting territory from the Hurons, who numbered perhaps one hundred thousand and previously controlled the St. Lawrence–Ottawa Rivers and the Ohio to Upper Mississippi River trade routes. After this successful conquest, the Iroquois nation expanded into the region of the St. Lawrence, where they met resistance from the mighty Algonquians. In the time of Jacques Cartier's three voyages up the St. Lawrence to Quebec and Montreal (1530s), the settlements had Algonquian names. But by the time Samuel de Champlain arrived (1603), those same towns had Iroquois names. But this was not to last. The Iroquois, in fact, were in the process of retreating and consolidating their small forces in the heartland of New York State, and neighboring tribes cheered when Champlain killed two Iroquois chiefs at Lake George in 1609. This event was inauspicious, casting a dark pall over French relations with the Iroquois for the remainder of the seventeenth century.

The Iroquois, constantly in a cycle of military advances and retreats with respect to neighboring tribes, became the geographically most significant North American group to be in contact with Europeans from vastly different backgrounds—English, French, and Dutch—who settled the various riverine arteries into the nascent American landscape (the St. Lawrence, Susquehanna, Allegheny, Delaware, and Hudson Rivers). They became dependent upon European trade goods—guns and powder from the Dutch, fishhooks, knives, and other metal from the French, alcohol from the English—and they created in the Europeans an equal dependence on the furs they could provide through subjugating their neighbors who had access to prime fur-trapping habitat. The Iroquois were ferocious in defense of their status as intermediaries. Trade was crucial to their long-term success.

Trade, however, then as now, is resource dependent. In order to have any sway over political affairs, the Iroquois knew that they had to produce fur for their European trading partners at Fort Orange and other posts down the Hudson, Delaware, and Susquehanna. Throughout the 1600s they brutalized their neighboring nations to the north (Hurons particularly), intercepted and robbed fur-laden Ottawa flotillas bound for European markets, and stole whatever they could from weaker neighbors because they could not produce enough fur themselves. The prime trapping habitats were all in Canada, lands occupied by Algonquian nations. French and English traders could go directly to these people for goods, but it was the desperate goal of the Iroquois to make sure that they had to pass through Iroquois lands to do so. Sometimes this meant expanding into territories of neighboring nations. Aggression between Indian nations was driven by the same motives as that between Indians and European powers. As we've seen throughout this book, resource limitation and habitat utilization are the underpinning of population wars.

By 1684, the English, at war in Europe with the French, could not provide documents to justify their occupation of the Iroquois territory. The French, however, had produced the treaty of 1624, signed by the Iroquois and Champlain, allowing France access to Iroquois territory.[112] Although they allowed passage of traders, in reality the Iroquois didn't want any permanent European presence in their land—they were hostile to any attempted posts—because such posts would eliminate them as middlemen.[113] Tribes or families who were able to find a position between groups—that is, middlemen—benefited throughout this period of history. The Iroquois recognized that there was often a lucrative space between two or more groups who—whether because of politics, location, or language—were unable to communicate or interact directly. They had to adapt as the population dynamics shifted. Political or cultural upheaval can be bloody and painful, but it can also create opportunities for populations who are willing to embrace change and take advantage of those opportunities. This is true across the entire spectrum of life, from humans to bacteria. The population that persists is the one that is willing to adapt quickly to a new equilibrium of coexistence. But sometimes adaptation comes with tragic loss of life.

The Indian tribes who lived in the Northeast have often been characterized as warlike. However, this is a misconception. Modern scholars believe that warfare was not as common or as deadly as the early historians suggested. A group known as the "Neutral nation" lived west of the Niagara River, grew tobacco, and enjoyed a land of plenty—fish, game, and boatloads of beaver. They also had a source of flint to supply the growing numbers of guns that made their way into the region from Fort Orange. This tribe remained neutral between the Iroquois to their east and the Hurons to their north, who hated the Iroquois. The Neutrals derived their wealth and prosperity from acting as middlemen between the suppliers of fur (Huron) and the buyers of fur and flints (Iroquois). They didn't dare trade directly with the French, for doing so would throw off the carefully established trade equilibrium set up by their suppliers, the mighty Huron nation. It would do them no good to fight with either Huron or Iroquois for direct access to European trade centers at Albany or Montreal. Like so many Indian nations, the Neutrals' prosperity and freedom were tied to trade relations and deeply ingrained traditions of conduct between producing nations and consumers. According to the historian George Hunt, we should not marvel at this peaceful situation, for it was common, not rare. Neutrality in trade was the norm in North America, rather than warfare:

> The Seneca, eastern neighbors of the Neutrals, were completely at peace with the Hurons until after 1639, and as they were the most numerous and most powerful of the Iroquois cantons, and as the Hurons outnumbered the Neutrals three to one, an assault upon either would have been a piece of unparalleled stupidity. . . . For either Hurons or Iroquois to have forced a belligerent status upon the Neutrals would have meant

> one more enemy for the nation foolish enough to do so. For the Neutrals to have gone to war would have meant participation in a terrible struggle which did not concern them. Therefore, they stayed home . . . traded their flint, their tobacco, and their furs to the Hurons, and lived at peace and at ease.[114]

In the 1600s the Algonquian tribes in Canada derived all their wealth from the French, who collected goods from all of Upper Canada and had free trade relations with numerous groups at Quebec, Tadoussac, and Montreal. The English and Dutch, meanwhile, received goods from Canada only by trading with intermediaries, principally the Iroquois, who controlled the valleys and waterways of upstate New York. By the middle of the seventeenth century the English took over Dutch Fort Orange, on the Hudson, while the French controlled the forts on the St. Lawrence drainage farther north. The Iroquois and Hurons ended up bitter rivals because each wanted to protect their exclusive trade relations with Europeans.

Hurons traded far and wide throughout the northern Great Lakes and protected their monopoly on goods produced by the Neutral nation and Petun (corn and tobacco producers) by preventing the French from going directly to them. As middlemen for furs from the West and the French in the East, as well as true middlemen with other tribes, the Huron were vulnerable in the 1600s. Their basic sustenance was all imported: Fish came from Nippissing who traveled to upper Lake Superior in winters, fur came from the Ottawa farther east, and corn came from the Neutrals and Petun. Over time Huron agricultural production waned. Since they depended so much on imports for their basic sustenance items, Huron starved by the thousands, while losing few in actual battles, when the Iroquois stormed in (in 1649) and took away their economic viability as direct suppliers to the French.

This disruption was economically and culturally motivated rather than a war of righteous ideology. The Iroquois wanted the riches that French ships brought from Europe, and they saw most of the best merchandise going to the Hurons. A lot of French traders, however, dealt directly with the Iroquois in the 1630s. Iroquois tried to establish peace treaties from time to time that outlined the terms of trade. The deadly skirmish of Champlain in 1609 was not a lasting strain on the trade relations between the French and Iroquois as much as it was an impediment to the good will and trust between the two nations. Necessity and angling for exclusivity drove much of the contact between Europeans and Native Americans. Furthermore, the French intervened in attempts by the Iroquois to negotiate peace with the Huron for fear that furs would pass through Iroquois to the Dutch at Fort Orange.[115]

There was much meddling. A French Jesuit missionary, Father Isaac Jogues, was sent to live among Mohawk Iroquois in 1646. His main goal was to persuade the Iroquois to forbid upper nations from using Iroquois lands as passage to Fort Orange so that the French could keep all the upper nations' goods for themselves. The Iroquois were most interested in trade, not bloodshed. If not for the French intervening for their own benefit, relations between Iroquois and other tribes might have reached an equilibrium perhaps more similar to that of the Neutrals and Petun with their surrounding nations.[116] The Iroquois raiding of trade canoes on the St. Lawrence River increased in 1641 and 1642, when they realized that Huron trade goods were going to the French, and the French were disproportionately meting out the best goods to nations other than the Iroquois. Inevitably the Dutch traded guns liberally with the Iroquois, and by 1649 the Iroquois took to annihilating the Hurons freely, out of their envy of Huron relations with the French.[117]

Hence the imagery of Native Americans as "warlike savages" is part of a false narrative, serving only to characterize their behavior as evil in order to justify an effort to root out and ultimately extinguish it—as if the vanquishing of a population might erase the existence of evil in the world. The correct narrative comes from the understanding that all populations are subject to basic environmental necessities. Humans are economic beings, which means that our actions at the population level are guided ultimately by resource utilization. When we meddle in economics through political or military actions, tragic consequences arise in the form of suffering and bloodshed.

The Iroquois traditionally took prisoners rather than kill all the members of enemy nations. By this tactic the Huron population became devoid of warriors soon after their defeat at the hands of the Iroquois. Despite this, however, the numerous wars waged by the Iroquois throughout the seventeenth century resulted in numerous losses to their warriors. Their Huron enemies, although originally a nomadic nation, were either killed or adopted into Iroquois families that had lost warriors in battle. Other Hurons dispersed and were incorporated into tribes farther west. Still others relocated around the growing towns of the St. Lawrence

River—Montreal, Quebec, and Trois-Rivières—where they were baptized and became known as French-Iroquois (later métis). Assimilation was the natural and intended result of warfare between the Iroquois and its "enemies." Warriors often had numerous wives, some from raids on other nations.

By the end of the seventeenth century, Algonquian and Iroquois, were tired of fighting each other, and the French wanted to facilitate friendly relations in the continuing interest of securing New France as a "land of humanity and peace in a world of cruelty and violence."[118] A tree was planted in Montreal during what came to be called the Tree of Peace Ceremony. This grand summer event was attended by roughly three thousand French and Indian men, women, and children from thirty-nine different nations. Montreal (founded in 1642) was the westernmost European settlement in the New World at the time. With roughly two thousand permanent inhabitants in 1701, it was a bustling town whose shops were filled each summer with visiting Indians from numerous nations eager to exchange their furs for metal goods, gunpowder, and hats. On this particular day, the host city for the ceremony saw Indians from both the Algonquian and Iroquoian nations and others assembled in droves from locales as distant as the Mississippi River. The governor-general of New France at this time was Louis-Hector de Callière, and he presided over the treaty negotiation alongside Philippe de Rigand Vaudreuil, the future governor of Canada. They sat regally perched on a vast open plain alongside the St. Lawrence River, where French officials had built a covered platform to receive all those who wished to speak. One after another, Indian leaders spoke, shared the calumet (peace pipe), and signed the treaty that was presented to them by Callière.

The "tree of peace" became a metaphor in the native culture of many nations who attended the treaty signing in Montreal on August 4, 1701. Shortly thereafter Vaudreuil secretly negotiated a peace between France and the English of New York, agreeing to refrain from attacking each other. In this way the shade from the tree of peace also covered New York.[119] The lasting importance of the Tree of Peace Ceremony has been summarized thus: "It was an incredibly resonant symbol and metaphor for coexistence. It structured the ways that Iroquois, French, British, and Algonquians spoke to one another in the colonial era and it remains to this day the symbol of the Haudenosaunee Confederacy.[120]

The relationships between the various Indian nations are a good example of a central truth about human population wars. The end result of war is never complete annihilation of the enemy. A population maybe temporarily subjugated, but it will rarely be destroyed—and it's foolish to think of this kind of destruction as the most desirable outcome. Instead it is wiser to accept the reality: that at some point in the future the so-called defeated population will remain as survivors and fellow citizens. The "enemy" does not exist in a vacuum. The best we can hope for is that the "victors" in a war might negotiate the terms of the future coexistence.

If a war is costly (in lives and money), then it seems most prudent to foster a mutually beneficial arrangement for all populations involved. The Iroquois did this by marrying the women of enemies, and adopting the defeated warriors and children into their tribes to replace their own dead warriors, a tradition called "mourning wars" or "mourning rituals."[121] By doing this they replenished and strengthened their own populations and limited the chances that their former enemy would try to destroy them once again. The upswing of the mourning wars was that tribal populations became amalgamations of individuals from a variety of ethnic backgrounds.[122]

It's a simple lesson and one that the Iroquois understood: Accept that you can't completely destroy your enemies. In practical modern terms we can use this strategy as a deterrent to warfare before the conflict ensues. I don't mean to romanticize the actions of the Iroquois; they were often vicious and cruel. However, I admire their grasp of the realities of population wars. The Iroquois understood the pointlessness of a scorched-earth approach against other Indian nations. Instead they adopted a harsh but pragmatic philosophy toward vanquished populations: Assimilate the members with potential value, kill the ones, perhaps, who are determined to kill you, and move forward from there. The dispersion of the Hurons, following the swift and unexpected attack of Iroquois on their towns in 1649, brought them into contact with many a neighboring nation. By 1650 the remaining Hurons who had not starved or succumbed to the elements or disease became incorporated into other tribes. The Huron people's lifestyle and livelihood were so changed that they were scarcely recognizable as a separate nation any longer. Rather they became an assimilated people distributed across a vast stretch of geography, melding their cultural heritage with those of the various tribes who accepted them.

The Neutrals likewise lost their homeland and were dispersed due to Iroquois raids around this time. Mohawk and Seneca joined forces, and six hundred of them stormed a Neutral town in 1651. Of the sixteen hundred inhabitants, most of the able-bodied fled in panic, while the old, the infants, and the infirm were killed on the spot. Eight hundred Neutrals retreated west to an island near present-day Green Bay, Wisconsin, and some of them went south to meld with the Catawba tribes in Carolina. The year 1652 was the final blow to the Neutral nation. They became amalgamated with their enemy tribes, just like those who once called themselves Huron.

In 1654 the Erie nation succumbed in a similar manner after a band of seven hundred Iroquois warriors attacked them in their principal town southeast of Lake Erie. Many were enslaved. Most, however, simply moved away to join other tribes, and the Erie, as a recognizable nation of their own, existed no longer. They melded with other Iroquois over time. Some were adopted in mourning rituals by families as far away as Montreal. Some formed new towns and became known as the Black Minqua of the upper Ohio River.

It's hard to visualize how small bands of Iroquois fighters—akin to guerrilla warriors—using tactics such as the pillaging of canoes filled with trade goods, or the raiding of towns, could have been such a thorn in the side of an empire as important as that of France. But we must remember that in the seventeenth century fur from North America was as valuable as any commodity we can imagine. France's ability to procure such goods, and trade them with other European countries that had posts in the New World (such as the Dutch or English, for example), or to transport them directly back to home ports in France, was the true measure of the nation's influence and political power. It seems strange that a simple commodity like beaver fur could destabilize and even destroy Indian nations and lead to seemingly endless bloodshed. However, we should bear in mind that future historians may look at our current geopolitical population wars and shake their heads in bewilderment that such a primitive and unsustainable source of energy as crude oil could cause such suffering and misery.

Reading the history of Iroquois warfare makes it clear why the region between the Finger Lakes of New York State and the prairies of Wisconsin and Illinois was so unpopulated by 1650, and remains so to this day. By the 1650s a great geographic divide was in place. The canoe-bearing tribes of the East, master traders and trappers (including the Iroquois) on the one hand, and the hunters and rice gatherers of the West who had fled their homelands in Michigan and the southern Lake Erie shore to settle in Wisconsin (including the Sauk and the "Fire Nation" Potowatomi) on the other. Between them lay the vast Michigan Peninsula and the Great Lakes shorelines leading to western New York State, all of which were virtually uninhabited. The tribes formerly living there dispersed because they feared incursions by the Iroquois from the East. Most who weren't assimilated into the Iroquois settled in Wisconsin by 1650.[123]

The region of upstate New York, specifically the Finger Lakes region, was a land of dense forest and wetlands, crisscrossed by moccasin trails and sparsely populated by humans during the seventeenth and eighteenth centuries. Not until the Erie Canal was opened in 1827 did a wave of settlement and industry flow through and begin to increase the human population of this region. The easy passage from the Hudson River to the Great Lakes afforded by the canal provided a new option for European immigrants: settlement of lands farther west, on the shores of the western Great Lakes. Some European immigrants stayed in New York, settling along the Erie Canal. But the vast majority bypassed the former Iroquois lands of the Finger Lakes and western New York, never leaving the boat until they reached the nascent cities of Cleveland, Detroit, Chicago, or Milwaukee. Meanwhile, hundreds of miles of forest, marsh, and grassland remained untouched between those places. It is therefore not hard to understand why today, despite having all the creature comforts and utilities of modern life, western New York remains one of the least populated regions in the eastern United States.

And yet, aside from the terror induced by Iroquois raiding parties of the previous century, there seems to be another, equally convincing, argument for why these areas south of the Great Lakes and the Michigan Peninsula became so depopulated: regional depletion of beaver. Upstate New York was devoid of commercial quantities of beaver by the 1640s, and the beaver populations of the Michigan Peninsula met this same fate by the 1670s. By then the vast majority of beaver trade was in the Illinois country (an Indian nation occupying most of present-day Illinois). Iroquois had to roam widely to procure beaver. The region around Green Bay, Wisconsin, was a major trading post for numerous nations from the North and South, with tens of thousands of Indians and traders making seasonal visits there. Most of the furs at these events came from Canada or Illinois. Farther east was Iroquoia, whose borders included Fort Orange on the Hudson, Montreal to the north, and the Delaware and Susquehanna Valleys to the south. Iroquois had better access to European ports (Montreal, New York City, Philadelphia, and the Maryland/Virginia ports), but without a stable supply of fur in their own territory, the privileged geographical position of the Iroquois was moot with respect to trade advantage. Hence fur to the Iroquois was analogous to our modern-day dependence on fossil fuels. The lack of domestic supply required them to travel far and wide in search of suppliers, and to deliver their goods to customers. The Iroquois homeland therefore had no sites that became major trade centers that would grow into major cities.

By 1677 the French—particularly during Réné-Robert de La Salle's expedition to the Mississippi River—built special boats to travel to the Indian trading posts (like the one at Green Bay) to load up with fur. No longer did the Indians have to travel with flotillas to the French; now the French came to the Indians. Thus the Iroquois could not intercept other tribes' flotillas. This era was "a new phase of traffic, the actual transportation of furs on a great scale" due to the French construction of cargo boats such as La Salle's *Griffon*, a forty-five-ton bark (three-masted sailing ship) launched near the Niagara River in 1679. By the 1680s the Iroquois were no longer supplying the majority of beaver pelts. Fully two-thirds of the supply going to France came from the Ottawa nation north of Montreal and Quebec.

A major geopolitical shift took place in 1665, when Louis XIV sent soldiers to New France (the Carignan-Salières Regiment), they defeated the Mohawk Iroquois in battle, and the rest of the Iroquois made peace with France. After 1667 the Iroquois split up, many going north to settle near Montreal and follow the French lifestyle to some degree, while the other groups followed English ways and settled towns farther south. The Mohawk came to be one of the groups that settled near French missions along the St. Lawrence River, where they were converted to Christianity by Jesuits and became autonomous residents of New France. The raids on villages now ceased for a time, and a period of prosperity in trade and commercial exploitation resulted. Unfortunately however, alcohol and disease had begun taking a heavy toll on the Indians of the New World.

It is hard to generalize about the political will of the Iroquois population. Early writers of American history observed that they were a nation in constant flux. A fundamental economic need drove their most obvious activity: they coveted the furs that the Huron traded to the French at Montreal and Three Rivers (Trois-Rivières). Their aboriginal political institutions were based on values unlike those we take for granted today; so judging them as seekers of peace or warlike "barbarians" in the 1600s and before can be done only through the lens of ethnocentrism. Since the casualties of their wars were generally few, and their customs in warfare included adoption and forced assimilation of prisoners, it's hard to paint them as barbaric. One thing is certain; they had a uniquely advantageous circumstance in the history of the New World.

In terms of military advantage, any European nation of the 1600s, looking to expand its empire in North America, would covet the Iroquois homeland. They occupied a position of strength through their military tactics, agile command of the water highways (rivers and lakes) of the Northeast and upper Midwest, and territorial control over the headwaters of the farthest reaching drainages on the continent (the Mississippi, Susquehanna, and St. Lawrence). From the major ports on the Eastern Seaboard—where all European nations entered North America—to the valuable interior headwaters, all nations had to pass through Iroquoia. Because of this, war between them and Europeans was inevitable based solely on the expansionist ethic that drove nearly all Western empires. But the French, who made it their goal to deal directly with the Iroquois in building New France, failed to occupy Iroquoia and failed to subdue or to exterminate them. The English and the Dutch fared no better in the early years of contact.

By the start of the eighteenth century the Iroquois had endured nearly two hundred years of contact with Europeans. They had persisted through countless rounds of trade, military skirmishes, and political engagements with foreign empires and neighboring tribes. Despite these challenges they were not in any meaningful way a colonized people.[124] And yet their home range was shrinking as European settlers encroached on its borders. Even though France was expelled from most of Iroquoia after 1760 (when British flags were hoisted over Montreal, marking the birth of modern Canada), what was to become the United States already had a citizenry active in horticulture, clearing forests, and settling the backcountry that previously was Iroquois hunting grounds. The settlers didn't understand that the Iroquois were, ecologically speaking, previously in harmonious equilibrium with the land, in the sense that they had reached a carrying capacity that sustained their population size.

Many scholars believe that the administrators of the empire back in Paris were just not dedicated enough to oversee the affairs of New France because they had more pressing concerns on the European continent. England, however, invested more of its manpower in military and emigration throughout the seventeenth and eighteenth centuries than did France. The settlers in the eighteenth century were predominantly Anglo-Saxon and, by the Revolutionary War, they found themselves fighting for the right to live peaceably on the land their grandparents had come to cultivate, under a new flag, and against the native Iroquois who had no "mother country" or European empire left to aid their defense.

The story of pre-Revolutionary New York is interesting both for its own sake, and for what it can teach us about just how delicate the balance can be between human populations. It's important to remember that despite some bloody interactions, there had been a long period in which the main participants more or less got along with one another in peaceful coexistence. When one of the first bloody encounters occurred, it nearly jeopardized the potential for friendly relations between the French and Iroquois for a century—the killing of two Iroquois chiefs by Champlain in 1609 mentioned earlier.

In the century and a half that followed, the lands around my house, like most of the American Indian homelands, were increasingly being visited, hunted, and settled by a mix of different ethnic and political groups from Europe. European settlers and frontiersmen, trappers and farmers, lived for many generations in homesteads that were scattered throughout Indian territory, interacting with the tribes and nations of Indians who had their own governments and expectations from political agreements. But there were other factions as well. Adventurous farmers and homesteaders from Germany, Sweden, France, Belgium, the Netherlands, and Luxembourg all contributed to the complex mosaic of human relations before and during the Revolutionary War. It's easy to forget that Europeans had been settling in this Indian backcountry for nearly two hundred years by the time Washington was inaugurated. The Indians, British, and various other settlers weren't necessarily friends, but they had achieved a sort of equilibrium with one another.

For decades prior, the English had had a more respectful relationship with the Iroquois in New York than did their French counterparts. In 1734 Britain named Sir William Johnson the superintendent of Indian affairs in New York. Johnson was an Irish emigrant who inherited a vast estate from his uncle in the Mohawk Valley near present-day Florida, New York. The behavior of Johnson (the most important representative of the British Empire in the region) toward the Indian natives shows how his political methods won the hearts and minds of the Iroquois. This excerpt from an important nineteenth-century book on the history of New York reveals clues of cultural assimilation that were highly significant in this period of history:

> He appears to have possessed the rare faculty of thoroughly adapting himself to surrounding circumstances, and . . . ingratiating himself in the favor of those with whom he was brought in contact. He could conform to all requirements, and was particularly happy in making himself beloved by all sorts of people. With his Dutch neighbors he would smoke his pipe and drink his flip [a kind of eggnog], as the incidents of frontier life or the prospects of the settlements were discussed, while, if occasion required, he could sustain his part in the most genteel company. With the Indians he was equally at home. He soon acquired their language and spoke it with great fluency. Their habits and peculiarities he studied, their wants he anticipated, and by a wise course he secured their confidence and an ascendancy over them which has scarcely a parallel in history. He is said to have possessed a hardy vigorous constitution, a strong coarse mind, unsusceptible to the finer feelings and "unconfined by those moral restraints which bridle men of tender conscience," he here saw the path open to wealth and distinction and determined to make the most of his opportunity. He often donned the Indian dress, out of compliment to his dusky friends, and at his mansion they were always welcome guests. Frequently when they came to consult him on some important matter, they made his house and grounds their home. He had on the Mohawk river two spacious residences known as Johnson Hall and Johnson Castle. Returning from their summer excursions and exchanging their furs for firearms and ammunition, the Indians used to spend several

> days at the castle when the family and domestics were at the Hall. There they were all liberally entertained by Sir William and 500 of them have been known, for nights together, after drinking pretty freely, to lie around him on the ground while he was the only white person in a house containing great quantities of everything that was to them valuable or desirable.[125]

What is left out of this account is the fact that Johnson fathered eight children with a woman, Molly (sometimes referred to as Mary) Brant,[126] a Mohawk Indian, whom he took as a common-law wife after the death of his first wife, a German woman named Catherine Weisenberg. Molly lived with Sir William at the palatial Johnson Hall, and they raised their children there together.

It should come as no surprise, then, due to a long tradition of shared respect and friendship instilled by Sir William Johnson—especially during the French and Indian War—that a significant faction of the Iroquois Confederacy sided with the British in the Revolutionary War.[127]

After William Johnson died in 1774, his son-in-law Guy Johnson (who was also Sir William's nephew) took over his estate and was eventually granted his father-in-law's superintendent position. Alas, Sir Guy lacked his uncle's subtle genius for connecting with the Iroquois. The Iroquois felt a strong sense of loyalty to the Johnson family, but Sir Guy did little to earn their continuing regard. His primary focus was on solidifying his family's power as the go-between for the British government and the Iroquois.

The most important character to emerge around this time was not a Johnson but rather Joseph Brant, a younger brother of Molly. Brant had been taken under the wing of Sir William and sent to school in Connecticut to live among Europeans and learn their ways. Afterward he returned to the Mohawk Valley to live near his sister at Johnson Hall, and work alongside Sir William. By doing so he garnered much respect among all Indians and impressed the British officials as well. Eventually he would become secretary to the British superintendent of Indian affairs.

Brant was a physically impressive and handsome man. There are several surviving oil paintings and drawings of him, probably because he presented such a striking and unusual figure. He dressed in a blend of European and Iroquois styles, and by all accounts he could move easily between the two worlds. When the Revolutionary War turned against the English and Iroquois alliance, he traveled to London in 1776 to plead the Iroquois case in front of the "Great Father," King George III. There he was received by sophisticated society and interviewed by the great British diarist (and Samuel Johnson's biographer) James Boswell. Brant seemed the perfect example of an assimilated man. He retained all the Indianhood of his Iroquois origins while accepting the knowledge and traditions of his adopted British culture. He was confident, bold, charismatic, and a passionate and wise leader. Brant had no hesitation in taking up arms against the Continental Army, detesting the city-dwelling colonists and many rural settlers alike. He was egged on by Sir Guy and his brother-in-law John (Sir William's son with Catherine), who displayed little of their father's humanistic spirit, restraint, and deliberating wisdom. Brant's excellent skill at oratory helped to garner much support in manpower and sentiment among his people for the British cause. It was he, after all, who would come face-to-face with Washington's dutiful Gen. John Sullivan in the opening battle of the bloody march through Iroquoia in 1779.

Brant grew increasingly resentful of the intrusions by the incipient American settlers and their appointed military officials. With increasing encouragement from the Johnsons, he traveled throughout Iroquoia as a go-between communicating the covert wishes of British Loyalists. The Johnson brothers—the cowards of this episode—later retreated to Canada when the fighting intensified, but continued to meddle through Brant and interfere with any attempts for peace between the British settlements of New York, the Indians, and the young American military.

The Indians, British, and American colonials met at various councils as the Revolutionary War picked up steam, and they all agreed that peace was preferable to war. One such meeting took place in Albany in 1775, another one in 1776 at German Flats, near Utica, and another one there as late as 1777. All these councils resulted in the sachems (tribal leaders) of the Six Nations proclaiming their desire for neutrality. The official resolution of the Indians, stated by the Mohawk sachem Little Abraham in 1775, was that the war between Britain and the colonies was a "family affair" and that therefore the Indians were content to sit by and watch them fight it out.

But neutrality was not easy to maintain in the face of bribery and promises of prosperity from the British. The Johnson family had such a powerful pull on the sentiments of the Iroquois that they could manipulate the entire Iroquois Confederacy through their loyal warrior-agent, Brant. The Johnsons were rich and influential among the Loyalist faction in North America, and they didn't hesitate to assert their wishes on the malleable emotions of the Iroquois.[128] Along with the promises of goods for all those who enlisted in the British army, a bounty was placed on every scalp brought in from the heads of settlers, colonials, and friends of the American Revolutionary cause.

The Battle of Oriskany took place in 1777 at a site just outside the bounds of today's Finger Lakes region, about twenty miles east of Lake Oneida. This battle was considered one of the bloodiest contests in the Revolutionary War, and it highlights the splintering effects that the war had on the native populations. One faction of the Iroquois nation, the Seneca, lured and tricked into battle by the English, became loyal to the Crown afterward. Shortly after the battle another faction, comprised of the Oneida and Tuscarora, became loyal to the colonials. In other words, the Iroquois Confederacy, after 1777, became engaged in its own civil war as a result of other populations at war, namely European and colonials.

The English accomplished much of their treachery via the ranger divisions, bands of English fighters who would today be called military "advisers" sent in to battle zones to organize fighters into formal armed divisions. The Seneca were told by English rangers to come and observe the battle at Oriskany and enjoy the rout, but they were not encouraged to fight. Almost all men from Seneca towns near the battle site went to watch what they hoped would be an English slaughter of the colonial army. Instead they found themselves targets. The colonial force, led by Gen. Nicholas Herkimer with eight hundred men, engaged the band of rangers, led by Capt. John Butler (the Johnsons' ally, in official military garb) and reinforced by Joseph Brant. Some of Herkimer's men were Oneida warriors, and the Seneca, perceived by their brethren as supporting the enemy English, had to fight for their lives. The great loss of life to the Seneca created a desire for revenge and, alongside the already allied Mohawk led by Joseph Brant, they drew deeper toward English loyalty in the Revolutionary War.[129]

The settlements of the Finger Lakes region were essentially hung out to dry during the American Revolutionary War. Indians, colonials, and Europeans, tried to go about their day-to-day business of raising families, producing goods, and getting along as neighbors. But the fear of battles, raids on villages, or the risk of being singled out as belonging to "the enemy" made life unbearable, forcing many to evacuate their homesteads and villages for safer havens in the larger cities to the east.[130] This rural backcountry, far away from the continental capitals, started out as a mélange of citizens primarily focused on the basics of subsistence farming, trapping, and survival. They would have followed the rituals of their Old World heritage, but they probably didn't have a lot of time to think about the bigger political picture that was about to engulf them. Instead they were pawns in a global struggle of imperial power that stretched across the Atlantic Ocean.

When the French were in control of New France, roughly 1600 to 1760, they were more focused on the major waterways and were content to leave this region as hunting grounds and sacred territory for the Indians, occasionally allowing their Jesuits to penetrate and set up shop. In the late eighteenth century, the British, like the French, focused on their forts along the waterways: Niagara on the passage between the lakes, Oswego on Lake Ontario, and Stanwix on the Mohawk/Hudson drainage. The British encouraged local production of goods, and the king sanctioned the building of roads and waterways to connect the interior towns, much more so than did the French. But a significant change happened in the Revolutionary War. By desperately desiring the Indians' assistance in their imperial affairs, they essentially abandoned their long-term imperial goal known as New England and America. After 150 years of intercourse between Europeans and natives, Britain turned its back on the settlers, and encouraged the Indians to brutalize settlements on the frontier that encroached on their turf.

British loyalists congregated around Fort Niagara, and Col. John Butler had a residence near the fort in 1777. He commanded a large body of British refugees, Tories who escaped or retreated from skirmishes farther east during the early years of the Revolutionary War. Joseph Brant fixed his residence at Lewiston, near Niagara, along with almost the entire population of the Mohawk tribe who had followed him there. From this location, on the border between Niagara Falls and the present-day city of Buffalo, Butler and Brant sent out predatory expeditions to the Mohawk and Susquehanna Valleys to pillage villages and murder European settlers who once lived in relative harmony with Indians in Iroquoia.

While concerted British military efforts were undertaken in 1777—to retake their now-occupied (by the Continental Army) stronghold in the Mohawk Valley, Fort Stanwix[131]—smaller raids were undertaken by Butler's and Brant's military expeditions. Now an official captain of the British military, Brant led his Mohawk warriors to sleeping villages and massacred women, children, and the elderly. Most of the fighting men of these towns had already shipped off to join Washington's Continental Army. British soldiers and Indian warriors joined forces to lay siege to helpless settlements. One of the most outrageous of these raids was known as the Wyoming massacre, near present-day Scranton, Pennsylvania, in 1778.

The British military brutally murdered English-speaking settler families simply because their settlements had been targeted for destruction. Some were murdered by their own family members who had taken off to join the Tory army. The Indians didn't fare much better in this respect. These settlers were often mixed families, the children of Indians who had married Europeans. Indians in the British military killed their own kind if they were found living in settler's villages. Brant and his men showed no mercy with their tomahawks and scalping knives. They followed an ancient tradition of brutalizing some of the defeated warriors and tortured them for days.

Such incidents illustrate a fundamental propensity of populations: They are thrust into upheaval at various stages of coexistence. When human populations commingle, a complex chain reaction ensues, resulting from geographic and environmental codependence, leading eventually to a mixing of ideology. A group of people who once shared core beliefs can be torn apart if some members adopt new ideas or beliefs about their place in the world. Some Iroquois warriors performed these grim deeds in part because they desperately wanted to prove their loyalty to the king and show that they were worthy of the gifts promised for their service. Others wanted revenge for losses in their tribes or families.

These kinds of backcountry hostilities were nothing new. They had been going on for decades as Europeans of German, Dutch, English, and Swedish origin settled lands bordering Iroquoia.[132] Settlers' lives were always dangerous; they knew that if the European and Indian cultures clashed, they might have to take up arms and fight as well. George Washington had a vested interest in the lands of the Ohio Valley, having surveyed and staked claims on huge tracts of that land as his own. Eventually he became the leader of a revolution, and would tolerate neither a British claim to control of those lands, nor Indian harassment of colonial settlers who might become productive tenants throughout the backcountry of the United States.

Washington's army was battered, poor, and of low morale by 1778. Soldiers were unpaid, and most had left their families alone, some in the wilderness villages now beset by murderous Indians spurred on by the English military. Washington decided that his troops needed rest, and that the fall and winter of 1778 would be a time of relative calm as he deliberated and meditated on what to do as a punishment for the Indian attacks. During this period the commander came up with a new campaign. He believed that an offensive in the heart of Indian territory—that region between Lake Ontario and the Susquehanna drainage west to Ohio—would destroy the Six Nations and their homeland, and send them running to seek shelter with their masters at Fort Niagara. This in turn would strain the English army and neutralize its dominion over the wilderness regions of the Western frontier.

On February 25, 1779, Congress gave a direction to their commander in chief: "to take effective measures for the protection of the inhabitants (of the western frontier in New York and Pennsylvania) and chastisement of the savages."[133] It is worth noting, however, that Washington knew very few specifics about who the inhabitants were. Maps at the time showed almost no detail about the towns of the interior in upstate western New York except that it was the heartland of the Iroquois, and that many settlers' villages existed there, which were beset by violence caused by the Indians.

Washington spent much of the winter of 1778 studying imprecise maps of Indian territory and reading reports that described Iroquoia. Orders were sent to General Horatio Gates to head the army for the expedition, but he refused, claiming that a younger man was needed to lead the war party (much to Washington's displeasure). The command was handed to General Sullivan. It was agreed that a five thousand-man army be assembled and led into the heart of Iroquoia. Each volunteer who joined the march would be given Continental Army rations and a one-hundred-dollar bounty per Indian.

Two thousand men would assemble in the southern part of New York's backcountry, along the present-day border of Pennsylvania at Tioga on the Susquehanna River drainage. Three thousand men would assemble in the north to block access along the Mohawk Valley, which drains to the east, toward Albany, thereby preventing a retreat of the enemy toward that city. The only path for the enemies' retreat from this five-thousand-man march into Iroquoia was west toward Fort Niagara, which could then be taken by force by the Continental Army and New York militia.

According to Washington, the goals of this expedition didn't even require a surrender at Fort Niagara. The mission was intended to "cut off those Indian nations, and to convince others that we have it in our power to carry the war into their own country, whenever they commence hostilities, it will be necessary that the blow should be sure and fatal, otherwise they will derive confidence from an ineffectual attempt, and become more insolent than before."[134] And furthermore the attack was to be waged against the Six Nations (Iroquois), their associates (British commanders and soldiers), and adherents (British Loyalists, and possibly French traders or translators as well). "The immediate object is their total destruction and devastation, and the capture of as many persons of every age and sex as possible. It will be essential to ruin their crops now in the ground, and prevent their planting more . . . and should be done in the most effectual manner, that the country may not be merely overrun, but destroyed." Washington told Sullivan that if the Iroquois were subdued and wanted to sue for peace, "You should encourage it, on condition that they will give some decisive evidence of their sincerity, by delivering up some of the principal instigators of their past hostilities into our hands—Butler, Brant, and the most mischievous of the Tories that have joined them."[135]

And so commenced the first extermination campaign against human beings in American history. The major battle of the Sullivan expedition took place about twenty miles south of where I am now writing at my desk, at Newtown (today we call it Elmira). During my research for this chapter, I happened to be on tour in England with my band. One bright, sunny August morning I popped into a London used bookshop. In the history section were a few titles on the history of New York. Paging through the book that I subsequently purchased, an 1885 commemorative volume on the history of the Sullivan campaign, I read some grisly journal entries that detailed the killing of Indians in the Battle of Newtown, and the skinning of dead battle victims in the woods only miles from my house.[136] It was a surreal moment for me, being in London and finding a very rare book that detailed little-known facts about the topography and physical features of the forests surrounding my home, the land with which I was so familiar. The geographical expanse of thirty-five hundred miles that separated me from my home on that day was analogous to the bewildering distance of disbelief that overcame me when I read the eyewitness accounts of atrocities that transpired there in historical times. These are tales not taught in American schools. I thought, Did I have to come to London to discover this?

Newtown sat on the bank of a major navigable river, the Chemung/Susquehanna, and it was a logical place for a battle. Boats of military supplies were shipped upstream from the Wyoming Valley (the home of present-day Scranton), where Sullivan had his headquarters in June 1779. After nearly three months of stagnation in Wyoming Valley—much of it spent waiting for basic supplies (a third of the fighting men who enlisted didn't even have shirts on their backs)—Sullivan and his army marched north and west and built a fort in late August at Tioga, New York, about twenty miles downstream from Newtown. On August 22, Gen. James Clinton of the northern Continental Army, whose troops numbered roughly three thousand, met Sullivan's army of nearly two thousand at this fort, later named Fort Sullivan.

With nearly five thousand men preparing to head into the heart of Iroquoia, British and Indian scouts were keen to pick a spot to engage the advancing American army. Newtown was just such a place. The British and Indian force had built breastworks that stretched more than a mile long and turned a settler's house into a fort along the line.

On August 29, 1779, General Sullivan's army marched toward Newtown and engaged the enemy. Among the 1,500 or so defenders of Iroquoia, roughly 250 were British military personnel, including the Johnsons, the Butlers (both General John and his son, Walter), and Joseph Brant, who led the roughly 1,200 Indian warriors into battle.

Drastically outnumbered, after a spirited fight lasting roughly six hours, the Indians and British fell back and scattered into the forest, retreating deeper into the Iroquois homeland. The tallies of the dead were less than twenty on the side of the Iroquois and British, and three on the American side. Wounded numbered less than fifty on both sides. During the retreat some wounded Indians were pursued by a detachment of Americans, and after about two miles they were left to die of their wounds sustained in battle. The vast majority of Iroquois and their British instigators withdrew northward, up the Catharine Valley toward

Seneca Lake, and by the next morning arrived at the largest Indian town in the region, Catharine's Town.

This was a pivotal moment in the history of America, but it's unlikely that the participants fully understood how the last two centuries had led to this moment. The British and Iroquois had retreated first to a town that perfectly symbolizes the complex history of this region, and whose most famous inhabitant symbolizes the diversity of the American Indians of this region at this juncture in history. The town was called Catharine's Town (sometimes called French Catharine's Town, or by its Iroquoian name, variously spelled Gasheoquago, Sheoquaga, or She-Qua-Ga). Today it's known as Montour Falls, New York. It was a clearing in the forest, roughly a hundred acres in size, sited along the flats of a rushing stream that today is prized and frequented by local fishermen. Much of the town was devoted to producing corn and beans, a surplus of which was supplied to, and subsidized by, the British to help with subsistence on the frontier.

Catharine's Town was named after Catharine Montour, a multilingual Seneca "queen" whose husband was the most important Seneca chief, Thomas Hudson (or his Iroquois name, Telenemut). Catharine married the chief when she reached womanhood, and together they kept and sold horses. Maintaining her command of French and simultaneously learning Iroquoian languages throughout her life, she was admired as an orator and for her captivating personality, and had a great influence on the affairs of the tribe. Before the Revolutionary War, she and her husband were often invited to attend important gatherings in the homes of Philadelphia's elite.

Catharine and her family were the end results of a complex story of cultural assimilation. The Montours were legendary among the French and Indians. They can be traced back to Isabel Montour, a woman whose parents were French (father) and Algonquian (mother). Such ancestry was fairly common in those days (a "race" known today as métis).[137] Isabel was born around Trois-Rivière into a family whose wealth was built on the fur trade, facilitating agreements between French, Iroquoian, and Algonquian merchants. The métis could communicate comfortably with all three of these cultures and were naturally suited to earning their confidence. Their cultural assimilation paid off; families such as the Montours lived in the most comfortable houses in town and wore the finest clothing, which resembled European high fashion mixed with Native American styles and accessories.

Isabel Montour, a child of this cultural milieu, experienced all the luxuries and creature comforts of the highest society in New France for her first ten years. She was kidnapped, however, at the age of ten by a raiding party of Iroquois warriors, adopted, and taken to live among them.

The early 1700s were a time of active trade between Europeans and Iroquois, and Isabel was seen as a useful intermediary as she grew up. Speaking both Iroquoian, French, and Algonquian tongues, she became a valuable interpreter and respected member of her tribe. Eventually she entered the highest levels of European colonial society when she was hired as the personal interpreter for the governor of New York, Robert Hunter, in 1710. This job, and her reputation as a gracious intermediary, ingratiated the Montour name to the Iroquois for generations to come. Her granddaughter, Catharine, born to Isabel's son, Andrew, continued the métis tradition of living in Indian villages and creating a culture of assimilation.

Having the genes of Algonquians and Europeans, and raised in the culture of the Iroquois with a strong command of French language and customs, the Montours and their heritage represent one of the best examples of human assimilation. Similar episodes of cultural and genetic mixing doubtlessly occurred time and again in the populating of the United States.

Historical records are vague, but the Montour descendants included daughters and sons, many of whom were important cultural go-betweens in Indian and European affairs. Serving equally important roles for Indian chiefs, European officials, and settlers alike, interpreters in the eighteenth-century were more than just linguistic specialists, they also acted as guides, advisers, escorts to important social gatherings, mediators, and negotiators. The Montour legacy is typical of many citizens of the North American population during this time who were neither purely European nor purely American Indian. It is a legacy that forms the core of the American identity.

Queen Catharine, the granddaughter of Isabel Montour, was greatly respected by her tribe, the Seneca. Catharine perpetuated a tradition of friendly relations between two cultures. She must have felt extreme dismay at the news of the advancing armies. As a true ambassador of goodwill between both European and American Indian culture, and as the inheritor of her father's deep wishes for his children to live freely in both the white and Indian worlds, Catharine was now in peril of losing her status and legacy at the hands of the Americans. Having already lost her husband in battle in a skirmish with southern tribes years before, and having sons off in battle elsewhere on the frontier, she fled in advance of Sullivan's army, and reached Fort Niagara, where she was well treated by the British.

In the hours before Sullivan and Clinton arrived at Catharine's Town on September 2, 1779, British military advisers deliberated with Indian sachems there about the best course of action to take against the advancing Americans. Some wanted to fight, others wanted to retreat to Niagara. Many Iroquois were fearful; they had lost faith in the "Great Father" (The king of England), and they no longer believed that the English would take care of them after the war. They knew that they were in a dangerous position, no longer truly supported by their allies and vulnerable to being destroyed by their enemies. The American army was on its way to erase all their villages—by burning the crops and houses—and kill anyone who stood in the way. The Indian chiefs and British officers decided, in advance of Sullivan's troops, that it was best to abandon Catharine's Town and all others along the way and proceed with the British army to Fort Niagara (some 140 miles northwest), where they would be under the protection of the British fort.

When Sullivan's army reached Catharine's Town they found it abandoned. The only remaining resident of the town was an elderly Cayuga Indian woman who spoke of the deliberations between the English officers and the sachems from many Indian nations that occurred there the night before. The woman could not walk; she was described as more than seventy years old. Here the historical records contradict themselves: Some say that Sullivan destroyed the village and burned every last house, but left a suitable hut for the woman and provisioned her with food for a lengthy subsistence. Other records, however, state that two soldiers crept back, locked the old woman (in this version, with a teenage girl) in the hut, and set it on fire.

Having sent the enemy on a routed retreat, Sullivan's passage through Iroquois country was a relatively easy march after the decisive battle of Newtown. This victory is a turning point in the Revolutionary War that is often overlooked. The Americans needed something to feel good about; their army was desperate in 1779. The soldiers' morale was wretchedly low, and they were poor, tattered, and worn down by the long, slow grind of war. Sullivan's campaign was the major effort of Washington's army that year. Had they failed in their mission, British confidence in controlling the western frontier would have swelled, and the British bond with the Indian stewards of Iroquoia would have been unbreakable.

In short, the "progress" of the American enterprise might have failed utterly if Sullivan's campaign had failed to reach its goal of subduing the British and Indians in the Iroquois backcountry. If the British had continued their incursions into the American backcountry at this time, the colonies might have been restricted to the cities of the Eastern Seaboard. Sullivan's expedition is credited with "opening up" an area soon to become the main thoroughfare for the growing nation.[138] It must be remembered that within fifty years of this campaign, the Erie Canal passed through the heartland of Iroquoia, bringing a flood of hopeful emigrants, a population as numerous as in any of the eastern cities, into the new nation's "Midwest" to form the American industrial heartland—cities like Chicago, Milwaukee, Cleveland, Detroit, Buffalo, and Rochester.

The Iroquois lost relatively few warriors in this battle, yet they still fled in terror of Sullivan's advancing army. They weren't necessarily frightened for their own lives, but for the lives of their families and the welfare of their communities. Washington's instructions to destroy the Iroquois homeland paid off. No crops were left standing; no house or hut escaped the torch of the American army. They extended their march throughout the Finger Lakes region, up the eastern shore of Seneca Lake, and westward to the Genesee Valley that leads to present-day Rochester.

By mid-September, Sullivan began his return march from the heart of Iroquoia. Having reached the most important Seneca villages in the Genesee Valley, Sullivan sent detachments southward on three separate warpaths, all to join up with his own troops at their original starting point near Fort Sullivan (Tioga, New York). On their southward journey the troops found and torched no fewer than thirty additional Indian villages and hundreds of acres of their cornfields and peach orchards (this was in addition to the destruction of between five and ten other villages, including Catharine's Town, destroyed on the northward march). Each one of the villages was abandoned by the time the detachments from Sullivan's army arrived. Barring one unexpected attack on one of the scouting parties, the entire expedition after Newtown was essentially free of casualties.

According to the historian A. T. Norton (1879), many of the towns Sullivan's party came upon had houses that were framed in a European manner, unlike the typical, traditional Iroquois "longhouse." The influence of European carpenters had by this time made its way deep into the interior of Indian territory. Although no mention was made of white settlers among these houses, Sullivan did encounter white captives on his march, brought into Iroquois country from raids on settlers' villages. Their forced assimilation ended as soon as Indian scouts brought the news of the Newtown battle. The Indians dispersed days in advance of Sullivan's party, leaving white captives and white children behind for Sullivan's men to discover in various stages of health.

When the expedition was over, in October 1779, Sullivan was congratulated formally in a letter from George Washington to the entire Continental Army, which read:

> The Commander-in-chief has now the pleasure of congratulating the army on the complete and full success of Maj. Gen. Sullivan, and the troops under his command, against the Seneca and other tribes of the Six Nations, as a just and necessary punishment for their wanton depredations, their unparalleled and innumerable cruelties, their deafness to all remonstrances and entreaty, and their perseverance in the most horrid acts of barbarity. Forty of their towns have been reduced to ashes, some of them large and commodious; that of the Genesee alone containing one hundred and twenty-eight houses. Their crops of corn have been entirely destroyed, which, by estimation it is said would have provided 160,000 bushels, besides large quantities of vegetables of various kinds. Their whole country has been overrun and laid waste, and they themselves compelled to place their security in a precipitate flight to the British fortress at Niagara. And the whole of this has been done with the loss of less than forty men on our part, including the killed, wounded, captured, and those who died natural deaths.[139]

Sullivan himself was content with the success of his expedition, believing that he and his men had covered "every creek and river and the whole country explored in search of Indian settlements . . . except one town situated near the Alleghany, about 58 miles (southwest of Genesee) there is not a single town left in the country of the Five Nations."[140]

The congratulations, however, proved premature. Now more embittered than ever at the loss of their villages, the Indians resumed raids on border towns. In their own summary of the Sullivan expedition, the Iroquois chiefs assured their British allies that they considered themselves very much an intact fighting force: "We do not look upon ourselves as defeated for we have never fought."[141] They vowed to exact revenge on the whites when the harsh winter was over.

By the next spring, 1780, many Iroquois refugees left the protection of the British fort at Niagara and set out across their former home range to resettle towns and villages that were destroyed.[142] Returning home after the war, Catharine continued her residency at a reinvigorated Catharine's Town until her death sometime in the early years of the nineteenth century. It is said that she met with Louis Philippe, the future French king, and a historical marker today commemorates his visit to Montour Falls in 1797.[143] Today Catharine's memorial grave site lies only a mile from the historical marker (her actual grave site is unknown). No roads lead to this spot tucked away in a deep forest clearing; It is only accessible by hiking trail. A moss-covered stone monument can be found there, engraved with Iroquoian symbols. The translation reads: "Every one of you always remember this." As moving as these words are, the memorial site itself is the most important thing symbolized here: Catharine's story is one of countless pieces of historical evidence, scattered throughout the American landscape, that demonstrate the complexity of cultural assimilation.

It's important to pause here for a moment and remember that this often forgotten episode of American history—the fighting in New York, 1779—represents perfectly the complexity and undeniable blending of traits seen in all population wars. All of the interactions between English and French empires in the New World involved nations or people whose heritage was not European. Many of the outstanding characters in this chapter, such as the Brants, Johnsons, and Montours, were of mixed parentage or were themselves parents of multiethnic families. It is furthermore well-known—although tangential to our story here—that at least one of our founding fathers had children with non-European mothers.[144]

These examples should not be seen as remarkable or unique. From the population perspective, they should be regarded as normal; the natural progression, if you will, of what happens to populations that come into contact with other populations. Traditional human warfare brings with it numerous unpleasant (and hopefully unnecessary in the future) side effects. But one unavoidable conclusion from this or any historical survey of human conflict is that the end result produces a commingled human population.

Sadly, the nineteenth century is also characterized by forced captivity of American Indians on reservations. In New York, most notably, is the large Iroquois community at Buffalo Creek (along today's Buffalo River) which began as a forty-nine-thousand-acre tract of Seneca land, settled by roughly one thousand refugees from the war. This community continued the traditions of their Iroquois people, holding annual councils around a fire where sachems from all across the land assembled to unify decisions for the national interest. Eventually this council fire was moved back to its original (and current) location on the land of the Onondaga people, just south of present-day Syracuse, New York. Many of the inhabitants of Buffalo Creek, however, were forced to relocate farther west in 1838 during President Andrew Jackson's Indian-removal program.

Not all Iroquois, however, were content to rebuild their old towns. Some had had enough fighting and moved to Canada, where Joseph Brant negotiated for a large land grant along the Grand River in Ontario. Today that land grant still remains, and a branch of the Iroquois nation, the Six Nations Reserve, lives there still. Other warriors, however, preferred to exact revenge on the Americans after Sullivan's campaign. These fighters became tools of the British, who used the seething resentment of the defeated Indians to fuel their imperial cause.

The Tories orchestrated raids throughout the backcountry and the Iroquois-British forces attacked towns in the Mohawk and Susquehanna Valleys. These raids destroyed roughly one thousand homes, one thousand barns, and six hundred thousand bushels of grain. The back-and-forth cycle of raid and retribution stopped only because the British had larger problems. The Revolutionary War was coming to an end due to events near the major cities farther east, which caused a withdrawal of Continental Army and British troops from Iroquoia. Finally, violence in the backcountry ceased for a time. Indeed, by the signing of the Treaty of Paris (1783), which limited the support that the British were allowed to offer the Iroquois, the Six Nations of Seneca, Mohawk, Cayuga, Oneida, Onondaga, and Tuscarora were essentially on their own. Unfortunately most of their heartland had been laid waste.

The ancient country of the Six Nations, the residence of their ancestors, from the time far beyond their earliest traditions, was included in the boundary granted to Americans.[145]

Human populations follow the same principle as those in nature: All populations inherit the fallout from their ancestors and create new relations among old templates. Populations are amalgamations of history and can only be appreciated in the context of that ancient narrative, lest we trivialize them and shortchange their significance. If, for instance, we think it is important to understand American history, then starting the tale at the Battle of Lexington seems an arbitrary decision. Just as trying to understand the causes of a flu epidemic is shortsighted without the knowledge of evolutionary history and the relationships of organisms involved in its transmission. The French were the first Europeans to make contact with the Indian nations in the Northeast. When the British pushed back the French, they became inheritors of an already long history of contact between Native Americans and Europeans, circumstances that were put in place by New France in the two centuries prior to the Treaty of Montreal in 1760. Instead of acting like good stewards of coexistence, the English often treated the Indians as enemies. When American colonists inherited this same attitude after the Revolutionary War, they proceeded to spend the next century attempting the futile act of extermination. Today we can look back in anger at such atrocities.

The original American Indian population is changed but not vanquished. The Six Nations are a larger population now than at any previous point in history. They take part in, and contribute to, many cultural norms that transcend any particular ethnic group, such as modern education, entertainment, and consumerism, activities that can be characterized simply as the modern industrialized way of life for all humans. In many instances American Indians blend this modernism with preservation and continuation of their original cultural traditions. Some nations have become wealthy from owning and operating casinos.

Current social problems of modern America, such as poverty and alcoholism, also afflict many people on reservations. These problems, however, are being countered by increases in available cash from gambling profits and other sources, money that is redistributed to build infrastructure, health-care facilities, cultural museums, and schools for the members of the tribal nation. More families are taking advantage of the American Indian College Fund, and other university programs that provide scholarships for Native American students. More universities than ever before have curricula devoted to Native American history and culture. Cornell University, for example, has an American Indian Program with its own devoted residence hall and administration building (the building name is Akwe:kon), purpose-built to celebrate American Indian heritage. Their mission is "to develop new generations of educated Native and non-Native peoples who will contemplate, study and contribute to the building of Nation and community in Native America.[146]

Although it's easy to argue that the American Indians as a whole are not equal in social status to many Caucasian subsets of our population, they are a far cry from being exterminated. In fact, suffering from poverty, disease, lack of education, and drug and alcohol addiction is on the increase for all groups of modern humans, and this is perhaps the most unfortunate result of population growth. If we care to only focus on one aspect of human life, the absolute number of suffering individuals, we encounter an inescapable truth, and one of the recurring themes of this book: As populations grow, the amount of conflict and suffering is inevitable. Therefore it should be our primary goal, as an assimilated group of humans arriving here from disparate historical trajectories and ethnic heritages, to alleviate the ravages of population wars as best we can in the interest of reducing suffering while increasing the longevity of our species. As a preliminary prerequisite to achieve these ends, the narrative of war, justified by the improper characterization of populations, must be changed.

8

COMPETITION IS UNTENABLE

Many narratives we use to explain our world are overly simplistic and therefore fundamentally flawed. Competition is one of them. We love the story that populations compete "to the death," as in warfare, with the stronger one wiping out the weaker. Yet, as we've seen with populations as diverse as pipe-dwelling bacteria and the Iroquoian people of North America, this simply isn't true. Populations are rarely eradicated; instead they assimilate with larger populations or find a niche in which they can retain many of their traits and continue their way of life in an equilibrium of coexistence.

The white settlers who entered Iroquoia came up with their own narratives to justify their intrusions, separate themselves from the Indians, and eventually force them from their own homeland. The tribes suffered but were not eradicated by the Europeans. Eventually the American Indians acted on their sovereign nationhood and began to participate in the economy and craft a new cultural relationship with modern society by establishing education programs, museums, and income-generating casinos on their land in order to improve their lives.

Human wars—even though we try to justify them with ideology—are comparable to natural population phenomena. The European settlers of the New World frontier and the American Indians were acting in a way that would be familiar to any evolutionary biologist or natural historian. They came into conflict as natural resources became increasingly limited, their home ranges expanded, and—after much bloodshed and fighting—eventually found a functional equilibrium that allowed the populations to coexist (this might be considered "dysfunctional," from the Indian perspective). Granted, all these populations were significantly changed by the conflict, but none of them was completely eradicated by it.

The story of early American history is not one of conquerors and vanquished people, it is one of evolutionary transformation. That is to say, from both a biological and a cultural perspective, assimilation of our species is evident. This assessment says nothing about the mistreatment of American Indians, or of their displacement, all of which should be viewed with compassion and sympathy. What we should take from it, however, is that the biased view—of a more "advanced" culture outcompeting or "rubbing out" a lesser one, is not tenable.

The understanding of assimilation requires a worldview that favors ongoing symbiosis over a definitive battle for existence. But that in itself is not enough. Another component is necessary: the environmental influence on the commingling populations. Such a perspective leads us to reconsider radically how we can coexist with other human populations and other species. To do so we will have to accept that many of the ideologies that we live by are flawed, and then replace them with more functional ones that allow us to engage with the populations that surround us while we try to alter the environment of our coexistence. Reassessing two dominant narratives—competition and free will—provides a starting point for such a revision. Both of these concepts have been misapplied to justify behavior in our modern life, while neither one by itself is sufficient to cause a new synthesis. I'll discuss the first of these ideas, that "life is a competition," in this chapter, and turn to the popular idea of "free will" in the next.

The idea of competitive struggle in populations began as a narrative device by Thomas Robert Malthus in 1798 to describe poverty and wealth in an increasingly industrialized society. His *An Essay on the Principle of Population, as It Affects the Future Improvement of Society*, published in 1798, treated food as a prize of sorts. His postulate was simple: As human numbers increase at an exponential rate, food production cannot keep pace, and therefore a struggle for existence results among all individuals.[147] Food, according to Malthus, is the primary limited resource, and famine is a natural impediment to (or "check" on) population growth. The human population, if left unchecked, would grow according to the exponential law of increase (x^2) while food can only increase at the much slower arithmetical law of increase (2x).

Famine was, to Malthus, an obvious limiting factor in the happiness and health of the human population. But it was also a constant threat to population growth. The more work expended in fighting off starvation, the less time and energy is spent on reproduction. The theoretical framework of Malthus's essay is straightforward enough, and makes perfect sense. It was, however, only the starting point for a much broader political polemic against the idea of charity for the poor.

The English Poor Laws were created during the reign of Elizabeth I (1533–1603) to help the poor survive by giving members of a parish money for food if they could not provide for themselves and their families. This money was raised by taxation on the middle- and upper-class citizens. By the time Malthus wrote his essay, this system had been in place for nearly two hundred years. He saw it as a perpetual drag on the upper classes of English society. It was the socioeconomic puzzle of dealing with the poor, rather than a serious intellectual consideration of competition, that motivated Malthus to write his essay. He believed that providing charity for the poor removes a natural check on the population (famine), therefore leaving them free to increase population size, putting those of the middle and upper classes (not to mention other poor people as well) in jeopardy of famine.

The solution, according to Malthus, was to abandon the Poor Laws because they did not erase poverty, and they put a strain on the upper classes to boot. He believed that policies to feed the poor essentially ignored the real problem of population increase (populations have intrinsic properties of increasing exponentially). The only change that could stem that increase was something "in the physical constitution of our nature"[148]—hence not policies, or political agendas, but rather personal actions and responsibility, principally "moral restraint," or birth control.[149] It was therefore imprudent, as Malthus saw it, to institute any policy that interfered with the natural competitiveness of workers. If they were assured of sustenance (from charity) what would motivate them to work? Hence competition between workers was a "necessary stimulus."[150] Without it there would be no industrial production in the lower classes, or checks on the growth of their population.

But something else depended heavily on competition being real: Malthus's reasoning. The very foundation of the Malthusian worldview would disappear if competition ceased to exist. For this reason, competition was not questioned deeply by Malthus, or by his most famous follower, Charles Darwin, as we shall see.

Malthus's essay was criticized for three decades after its publication, yet competition remained out of the spotlight. Instead critics focused on things like future improvement. They couldn't understand how an increase in population could be viewed as a threat to humankind. Intellectuals of the early nineteenth century believed that population growth must be a sign of prosperity, not limitation or danger. They were blind, perhaps, to the broad causes of poverty, and of course, the ones doing the writing in those days were of the privileged classes, trying to avoid responsibility for taking care of unemployed workers.

Despite its early criticism, the gist of Malthus's essay has come to be an often-repeated explanation for why some people have so little, while others have so much. Competition has hardened into a seldom-questioned truism. Ignoring historical factors, or the constraints imparted by socio-economics, most people today believe that the workforce is seen as a competition in which the winners get rewarded and the losers deserve their low status. In this over simplistic view of things, poor people have no reason to claim unfairness because the competitive struggle—the "game," if you will—is not rigged. Each member of society is in the same competitive arena. The feast-or-famine rules are the same for rich and poor alike: If you lose too many games, by not performing well in your job, you will sink progressively lower on the ladder of success. Those at the bottom are simply not trying hard enough, and they should be able to wrest themselves out of poverty if they would simply try harder, perhaps work more hours, or try some other job that's better suited to their abilities.

This grim view of society is not one to which I subscribe. I see diversity and variation in wealth and class as part of the process of mixing in our modern culture. I'm reminded of this every time I walk the streets of New York City, or any other major city for that matter. I am amazed at the diversity of our species. Contrary to the popular view of a monotypic, gritty, uptight, angry populace, New Yorkers are a kaleidoscopic conglomerate of cultures and physical appearances. Every face that passes seems familiar in some ways but mysteriously foreign in others. I try to guess the backgrounds or ethnicity of many whom I see, but I know I'm usually wrong. I'm biased, like everyone else, by my education and stereotypic thinking, no matter how open-minded I try to be. I have to remind myself constantly that supposedly "conquered" races and forgotten foreign customs are nearly as common here as are Midwestern tourists or New Jersey housewives. Who we consider as foreign is usually just a matter of what version of history is most familiar to us. The more I travel and the more I think about it, the entire concept of "foreigner" is becoming passé. Our species is amalgamating before my very eyes.

New York—like most big cities—is built on the idea that the most worthy of us win the biggest rewards. The hierarchy of success is all around us—I take the bus or subway, you take a cab, she takes a limo—you can't miss it even if you want to. No matter what you've achieved, or how much you have, it's easy to feel like a failure in the Big Apple. There's always someone doing better than you—or at least appearing to. Of course success is often subjective and sometimes short-lived. There's always some fitter, "worthier" person coming up from behind. Someone who, through diligent study and hard work alone, is now somehow better—faster, smarter, more creative, or more aggressive—than his or her predecessors. Our culture is built on the idea that life is a competition, and that the winners are more deserving of its rewards than are the losers. This idea crept into our daily lives from nearly constant exposure to sports, celebrity culture, big business, warfare, nation building, and relationships, among other aspects of modern society. It is now so ingrained in our collective consciousness that it is generally believed to be unassailable.

Perhaps it's because I've deluded myself into believing the "outdated" view of America as the land of equal opportunity for all, but I think this is nonsense. It seems that there are no immediate differences between me and those others scurrying about. New York City at rush hour is a good example of a level playing field. The vast majority of us are trying to grab the same taxi, or pushing through the turnstiles in a hurry to catch the next subway train. Some of us can hail cabs better—I can often "outcompete" less-able-bodied people on the same corner because I will walk into the street to make my presence known—but I've never been so desperate for success that I would take a cab away from an older person, or a mother with kids, for instance. The people taking limos may appear better off at first blush, but the fact is that they have simply opted for comfort rather than speed, because walking is often faster on the crosstown streets. If getting from point A to point B is the only measure of success, I might feel a momentary triumph when a cab finally stops to pick me up instead of stopping at the next corner to pick up someone else. On any numbered street in midtown, a fifteen-minute walk may just turn into a half-hour sit-and-wait in the backseat. The other commuters and I may have important historical or socioeconomic differences, but in this very moment it seems that luck, rather than skill, plays the biggest role in who succeeds and who doesn't.

We've been led to believe that competition is a dominating fact of success in life. The accepted dogma is that in our society you have to work hard to be the best or someone else will take a job or a resource away from you. Competing businesses will result in better goods and services for all.[151] The free market is king. Parents tell their kids to "do your best" as they head out the door to school. But most parents really mean to say, "Do better than the others" because they've bought into the seductive belief that privilege is earned by winning, and winning is the only form of success.

But is success in human life measureable by the same criteria we use to gauge success in biological competition among wild species? What about games and sports? Should we consider a person's lifetime of activities some sort of win-loss statistic? Popular notions such as "Life is a game," "Nice guys finish last," or "Coffee is for closers," all imply that rewards in life are easily measured and that success comes only from winning. We can thank Charles Darwin and the intellectual revolution he ushered into the minds of Western civilization for the notion that rewards in life can be boiled down to one thing: advantageous possession of traits that allow one individual to outcompete another in the struggle for life.

Darwin read Malthus's essay around 1838, after spending many years thinking about an explanation for the diverse adaptations and distributions of plants and animals that he observed on his five-year voyage around the world (on board HMS *Beagle*, 1831–36). Seeing the spoils of competition as "favourable variations," Darwin used Malthusian reasoning to suggest that under the conditions of a population's struggle for existence, favorable variations (those that led to success in survival and reproduction) would tend to be preserved, while unfavorable ones (those that failed to enhance survival or reproduction) would quickly die out.

The continuation of this process, through multiple generations, according to Darwin, would lead to new species, natural selection "picking the most profitable."[152] Hence all new varieties of species are spurred on by incessant competition. Because of the inherent properties of population increase, and the scarcity of resources caused by Malthusian principles, even a slight variation in a trait that enhances survival and reproduction can be considered a "win" for the competitors.[153]

Darwin's fame came roughly twenty years later, after the publication of *On the Origin of Species.* The popularity of his controversial book—which stated that natural law (competition and natural selection), rather than divine purpose, was responsible for the creation of new species, including humans—can hardly be overstated. The initial printing of twelve hundred books sold out two days after their release date (November 24, 1859). The second printing of three thousand books sold out the very next month. Without ceasing, the publisher released more copies for sale, and they kept selling out for decades to come. Darwin's science was not easily understood by most people, but his fame was established early on by the incessant lampooning of him as the originator of the idea that humans were descended from apes. His face appeared repeatedly in the mainstream weekly press, *Punch* magazine, *Vanity Fair, Harper's Weekly*, and *Fun*, and their equivalent publications in other countries. Darwin's ideas were discussed in universities and households all over Europe, Russia, and America, most people discounting or disbelieving his conclusions. Nonetheless, at his death, in 1882, Charles Darwin was buried at Westminster Abbey in London, alongside some of the most important people in English history—kings and queens, scientists and poets, statesmen and scholars—as a lasting testament to his indelible fame.

It was Darwin's idea, rather than his personality or actions, that drove his renown. While it is well known that the idea of natural selection originated with insights from Malthus, less well known is the fact that Darwin helped to make Malthus famous in return. As Darwin's natural selection became the primary mechanism for explaining evolution throughout the twentieth century, Malthus's ideas about competition came along with it. Together these ideas hardened into the core of greatest influence on most biologists' thinking. This happened long after both men were dead.

Today we have political and economic theory being informed by what happens in nature—natural selection operating among competing businesses for instance—and biology being informed by economics (ecology is often referred to in terms of "budgets" and the "balance" of nature). This tradition isn't new; it can be traced back to Darwin's time, when political economists began to adopt Darwinian ideas into their teaching.[154] The point here is that the tradition of using ideas from nature and applying them to economics, and vice versa, is an old one. Sometimes entrenched intellectual traditions like this operate without anyone challenging the logic necessary for their operation. Competition is such an important explanatory concept that a crack in its foundation might bring down the entire edifice of modern economics, political science, and modern biology all at once. The most popular narratives from all these fields have transcended the ivory tower and have taken root among millions of readers of award-winning books such as *Freakonomics*, *The Selfish Gene*, or *Civilization: The West and the Rest*. All these depend heavily on the idea that competition drives everything forward in an unending natural progression. Without competition, there is no progress. Without competition, there is no continuity of the intellectual tradition that links nature to socioeconomics. So it goes unchallenged.

Without questioning competition, however, we are susceptible to scandal. Suppose the traditional unchallenged concept of competition—as the stimulus and primary mechanism of natural selection—is wrong and is used instead simply as an easy excuse to perpetuate a deeply unfair social system by giving it a quasi-Darwinian spin: The only way to escape poverty, for instance, is through outcompeting your neighbor in some low-wage occupation (as opposed to associating yourself with institutions that provide living wages, and helping each other maintain that institution). Darwin wrote "Each new species is produced . . . by having some advantage over those with which it comes into competition; and the consequent extinction of the less favored forms almost invariably follows."[155] This idea, that the less-favored forms almost "invariably" go extinct has been shown to be incorrect. In the latter half of the twentieth century, much was written about extinction, and the discovery of the "Big Five" mass extinction events turned out to reveal something profound. Extinction was not caused by poorly adapted traits—in fact most forms seemed well adapted to their habitats. The problem was that populations could not change rapidly enough to contend with the extreme swings in environmental conditions.[156] This made scientists—as it should make us too—question the efficacy of competition and natural selection in sustaining the evolutionary process.

I try to minimize the importance of competition even when I'm actively engaged in it. In a sense this strategy minimizes the significance of winning, but it also makes losing feel more hopeful, like part of the process of life—not an ending but part of a larger continuum. Like most people raised in this society, I love to compete. My brother and I used to play games incessantly with friends in the neighborhood throughout our childhood. When we weren't arranging teams for touch football or baseball at the local schoolyard, we were designing shooting ranges in the basement for our BB guns, or playing endless rounds of Ping-Pong during cold winter days when it was too cold to throw the ball around outdoors. The fun I had in those childhood years, and throughout my teens, made me learn to love skilled activities. Nowadays I'm very competitive at sports, academics, and music. I've always believed, for instance, that my band is the best punk band in the world, and am constantly seeking to prove it. I take it personally if a particular show we've played gets a bad review. I'm also a hockey player in a local adult league. I'm one of the oldest players now; the other guys are younger and faster, but that doesn't diminish the pleasure I take in playing. They may play a better game than I do, but I still enjoy the chance to compete against them.

These activities help me recognize that there is more to competition than winning. I'm a huge advocate of sports, but I always knew that I could never play at the professional level in any of them. It doesn't matter to me. Participating in them at any level adds interest and health to my life. The pleasure is to play, even if sometimes you don't win. There's a thrill to be gained by just participating in the process. While it's true that each game is an opportunity to improve on what you've learned in the previous effort, it's really the feeling that you've given it your best, and had some unexpected surprises along the way, that makes sports so rewarding. And that's an idea that I think has been lost in our culture; the win-loss statistic is not an ultimate gauge by which to measure one's life.

Don't get me wrong. I still believe that only one team, or one individual, should walk home with a winner's trophy. I've heard that there are leagues for kids in some places that award trophies for everyone, regardless of skill, with the tacit suggestion that every kid is a winner.

This is a misguided attempt to encourage them to stick with the sport, perhaps, or simply to lie to them because some parents are afraid that their kids will become depressed if they're ever allowed to believe that they lack talent. This tactic puts an undue emphasis on the prize and actually detracts from the important functions of sports, such as skill development and social engagement.

Competition, as a concept that pits two individuals (or teams, or groups) against each other in the quest for a clearly defined goal, belongs on the playing field, not in the ideology of a nation or in the foundation of an ethical principle. Yet, as we read every day in the newspapers, competition is supposed to sustain our markets and our culture, for without it there would be no progress.

It's not hard to remember that the misapplication of the idea of competition nearly crippled our economy in 2007. In the first years of the twenty-first century, mortgages became incredibly cheap and easy to secure. Suddenly home ownership seemed like a no-risk deal, a guaranteed return on an easy investment. Would-be home owners in hot markets—Los Angeles, Phoenix, Miami—began to buy up properties. Owning a home—and making a big profit on that home—was so easily attainable a goal that only a complete loser wouldn't have his hand in the game and his signature on a subprime mortgage. If home owners had understood how completely corrupt the mortgage game was, they might not have felt the compulsion to jump into it. But once the idea of home ownership became competitive, all reason went out the window as the ideology of competition matured and hardened. When the available properties began to dry up and questions about real value came up, competition and panic set in—both in the bankers writing the loans and in the minds of would-be home owners. Competition—the need to keep up and hopefully best your peers—ultimately led to the implosion of the housing and mortgage markets.

There is a depressing variety of other examples of business competition having negative effects: leveraged buyouts gutting businesses and selling them for parts, setting the workers free without pay or compensation. The fossil-fuel industry, trampling over environmental concerns, is notorious for leaving a wake of toxic chemicals and debris behind to poison local communities. The ideas that more is always better, and the push for lower prices by any means necessary, is a constant mantra in our consumerist society. On a logical level most people must realize that this kind of culture is unsustainable and ultimately destructive—but it seems impossible to stop because the narrative of competition is so deeply engrained. Based on this foregone conclusion, all transgressions can be explained as a result of the incessant competition in nature.

I am preoccupied by this popular idea of competition, and what it means to us as individuals and as a culture. I am continually frustrated by the way that evolutionary science is mangled and contorted to suit the narrative that human life is some sort of playing field where only the most able and deserving competitors win privilege, wealth, esteem, or happiness. What's worse perhaps is the faux-Darwinian belief that you can do whatever it takes to claw your way to the top rung of the ladder of success, and your ascendancy is somehow "natural" and therefore "right." These harsh and damaging notions have no basis in evolutionary science, but rather they stem from the timeless belief in human progress.

Progress in evolution was a topic that was once very popular, but today it is a thing of the past. At the turn of the twentieth century (1900), Charles Darwin had been dead for eighteen years. Although his work in evolution had established the fact of descent with modification (or change through time of organisms along ancestral lineages), the so-called mechanism of this change, natural selection, was not well received by the vast majority of biologists. Part of the reason for the rejection was that Darwin had presented so few examples of natural selection in the wild in *On the Origin of Species*.[157] Another problem with the acceptance of natural selection at that time was that there was no understanding of why traits in offspring are so similar to their parents. Without understanding heredity, critics could claim that adaptations came from any number of supernatural sources such as intelligent design, use and disuse of parts (Lamarckism), or conformity to some preexisting archetype in the scala naturae.

Jean-Baptiste Lamarck (1744–1829) was a French evolutionist, predating Darwin, who argued that use and disuse of organs might be the cause for evolution. The classic Lamarckian example is that the use of a neck in giraffes for obtaining high leaves drove the adaptation forward, allowing long-necked individuals who strove for the highest branches to pass along this trait to their offspring. The Scala Naturae, sometimes called the "ladder of life," is a preevolutionary classification system based on the belief that species were defined by archetypes—ideal forms that were specially created in the mind of God. All archetypes could be arranged from low (worms and vermin) to high (humankind). Every organism could be placed somewhere neatly along the scale, in correspondence with its archetype. The scale was refined throughout the eighteenth century as more organisms were discovered, but humankind never lost its place at the top. None of the positions along the scale blended into others. Each rung of the ladder was distinct, and all species were assumed to be fixed and unchangeable.

The theory of heredity that we take so for granted today—based on the existence of DNA and the transmission of genes—was not incorporated into evolutionary theory until long after Darwin's death. Gregor Mendel, an Austrian monk who experimented with heredity in pea plants, discovered the secret of particulate inheritance (transmission of genes) essentially in an intellectual vacuum. From his garden plot at a monastery in Brno (in the present-day Czech Republic), he experimented with more than twenty-nine thousand pea plants in the 1850s and 1860s. He discovered trends in lines of crossbred peas—consistencies in the frequencies of traits such as seed shape, pod shape, and flower color—that led to the foundational principles of genetics (the law of independent assortment and the law of segregation). Although Mendel published his work in 1866, during Darwin's lifetime, Darwin did not read the journal of the Brno Natural History Society (written in German) and remained unaware of the work for the rest of his life. Others interested in evolution at the time of its publication ignored it because the findings were presented more along the lines of a paper on hybridization rather than heredity.

Around the year 1900 Mendel's papers were finally introduced formally into the discussion of evolution by researchers in various European countries—Hugo de Vries in the Netherland, Erich Tschermak in Austria, and Karl Correns in Germany. Each published independent findings on heredity, referring to, and inspired by, the work of Mendel. Now the field of evolutionary biology was poised to inflate the importance of natural selection as a law of nature that could distinguish between variant forms of hereditary materials. This marked the birth of the neo-Darwinians, those evolutionary biologists who put an overt emphasis on natural selection as the primary driver of evolution.[158]

Darwin would have found it remarkable that his name was attached to such a view as this. His theory of heredity, finally published as a provisional hypothesis in 1868,[159] was called pangenesis, and it was anything but particulate inheritance. It relied on blended molecules called "gemmules," for which there was no evidence, but these particles were a milieu from millions of sources within the body. Gemmules circulated throughout the body, picking up even smaller molecules along the way. Each organ was said to "cast off" gemmules (in Darwin's words). Every cell in the body gave off gemmules that eventually came to rest in the eggs and sperm. The sexual act, then, blended and transmitted gemmules from both parents. Offspring were thus the result of a blending inheritance, a mixture of traits from the gemmules of the father and those of the mother. Furthermore, these traits might be affected by the environment because the gemmules would come from many cells that came into contact with the surroundings of the parents. What's more, certain traits were enhanced by greater use. More gemmules would be created by organs that were used often in life while disuse led to degenerate organs.

Reading the passage on gemmules from Darwin's provisional theory of pangenesis is eye-opening because it is written with the same style and air of authority as many of his most important observations. But it is all speculation and imaginative, with no basis in what we now know to be fact. No evidence has ever been found that supports the existence of Darwin's gemmules or anything like them. Today we know that the germ cells (sperm and egg) carry information (in DNA) that will be passed along to every cell in the body. There is a complete separation in function and formation of the somatic cells (body cells) and the germ cells. Blending between the two does not occur.

The division of germ cells and somatic cells was known shortly after Darwin's death, and it was due in large part to the work of August Weismann in Germany during the 1890s. His germ plasm theory claimed that all the hereditary material is transmitted only in the sex cells, which are impervious to the effects of the environment or of use and disuse. In 1896 Weismann proposed his germinal selection theory, which was the founding spirit of neo-Darwinism. In this view, natural selection ultimately distinguishes between variant forms of hereditary material (good genes vs. bad genes, in today's parlance), and, through generations of hereditary descent (passing along those good genes), produces lineages that show progressive specialization and perfection. This, then, was the explanation for the well-known phenomenon of increasing adaptive perfection—for instance, the excellent vision of the hawk is an adaptation built upon the less-well-developed eyes of its ancestors; or the large brain size of humans is a progressive development of the less-well-developed brain of our ancestors (apes). In each case natural selection seems to be working toward a goal (excellent vision, large brain) that conveniently seems to lead toward the evolution of humans. Progress, therefore, appeared to be well supported by evolutionary science, and competition was justifiably viewed as a stimulus to progress.

So, around 1900, the neo-Darwinian worldview was set for the twentieth century: Natural selection was a law-like process that affected the germ cells of every sexually reproducing organism. By this process alone, all varieties of traits could be produced, and eventually perfected. Progress was an inherent feature of natural selection. Nowhere, however, in this developing worldview was there a serious consideration of the original Malthusian principle that provided the impetus for Darwin's (and co-discoverer Alfred Russel Wallace's) discovery of natural selection. Although Weismann had "preserved the actuality of a struggle and a selection,"[160] he had not questioned the efficacy of competition in driving the process forward, nor had any other neo-Darwinian of the early twentieth century. It was a necessary ingredient as the neo-Darwinian engine of evolution.

There were voices of dissent. For instance, the Stanford University entomologist Vernon Kellogg and his partner, Ruby Bell, published a treatise on ladybugs and other insects in 1904.[161] They showed that the number of spots on the most common American species, *Hippodamia convergens* (the convergent lady beetle), varies significantly between zero and eighteen, with most possessing twelve spots. Furthermore, they found no fewer than eighty-four different patterns of spot formation, and no detriment to successful reproduction for any of the patterns. This finding did not correspond to a stringent natural selection. The trait was highly variable, indicating that competition, as reflected by variation in spots and patterns, was not very strong, or possibly nonexistent. Certainly the struggle-for-existence analogy—individuals fighting with one another for some reproductive advantage with respect to their color pattern—did not seem applicable.

To counter those who saw natural selection as the primary mechanism of evolution (selectionists, as Weismann and other neo-Darwinians were sometimes called), Kellogg and Bell asked why such a noticeable trait, easily spotted by bird predators, would not be tightly constrained by natural selection. If twelve spots in a particular arrangement was the average, why was such an abundance of other varieties in the population being perpetuated with impunity each generation? Their answer was that natural selection was not all-powerful, competition was not that significant, and the ladybugs (and various other insect species in their study, which showed a similar hereditary tendency toward highly variable, obvious traits) got along fine regardless of their most conspicuous variation. Where is the progress if having only six spots allows one to live as well as those with eighteen? To them, and to a relatively small dissenting choir of nonselectionist Darwinians around the turn of the century, many traits of species offer no selective advantage, and therefore render the concept of competition moot.

Although not appreciated fully at the time of its publication (1904), Kellogg and Bell's discovery shows that natural selection is only part of the story, and that concepts of progress aren't universally supported by evolutionary considerations. Many traits, even obvious ones, seem to be selectively neutral, meaning that no selective mechanism is in place to weed them out. They somehow avoid natural selection. Traits that have no selective value are like competitors in a contest who aren't interested in the prize (like me when I play organized sports; I'm just participating and contributing to my own and my team's enjoyment of the game). Despite this challenge to the importance of competition and to the efficacy of natural selection, the neo-Darwinian worldview—that natural selection acts ultimately on the hereditary material to produce all traits—was already well-established in the first decades of the twentieth century, and it was on its way to becoming even more popular.

The 1920s, 1930s, and 1940s saw an increasing sophistication of evolutionary science from the application of mathematical approaches to heredity and natural selection. Mendelian inheritance supplied the variables (different varieties of genes), natural selection supplied the adaptive mechanism. Both were subject to mathematical treatments that produced satisfying models for the evolutionary process. During this time, evolution became characterized as a "change in the genetic composition of populations."[162] Natural selection could now be explained mathematically, using relatively few variables, as nonrandom, differential survival and reproduction of individuals (reflected by the genes they carry). Theorists had worked hard during these decades of the twentieth century to whittle down the list of variables considered important in the evolutionary process.[163] Competition, however, still formed the underlying assumption that drove this process forward.

This period of intellectual development in evolutionary biology is called the "evolutionary synthesis,"[164] and it fostered numerous developments, including the rise of the field of population genetics (concerned with the quantification and modeling of a population's genetic makeup); the discovery of small mutations as contributors to a population's normal range of variation; and, important for our discussion, the rejection of progressive and purposive explanations for evolution. The restricted list of variables considered important for evolution were things like mutation rates, size of breeding population, amount of genetic variation, and so on. All these things were subject to experimentation, and were malleable to some degree in laboratory conditions. The field of evolutionary biology turned a corner, and researchers focused less on trying to find evidence of purpose or progress in nature, and more on which empirical variables were the most important in the origin of species.

Through this period and into the 1960s, the neo-Darwinian approach was the only game in town. With all the data coming in from new discoveries about DNA and genes, natural selection as a sole, unwavering mechanism that distinguished between slightly favorable and slightly deleterious mutations served the mathematical models well. In turn, these models produced satisfying results that verified the effectiveness of natural selection in producing adaptive traits. One would think that something critical of competition might come out of this period. But in fact, just the opposite occurred. Competition hardened as the underlying mechanism of natural selection. All other theories of evolution were eliminated from the table of discussion, leaving only mutation and natural selection as the mechanisms of change through time. Competition now became even more important. For without slight variations competing for some adaptive purpose, what is selection acting on?

Other studies began to surface, however, that focused on the widespread abundance of trivial characters—traits passed from parents to offspring but clearly not contributing to any adaptive purpose. For example, two bird species that are closely related, and overlap in geographic range, are the song sparrow (*Melospiza melodia*) and the swamp sparrow (*Melospiza georgiana*) of eastern North America. Although they have slightly different habits, they are very difficult to tell apart. Individuals from either species show similar feather markings, a dark eye stripe, two facial stripes below the eye, and white feathers on the throat. One of the only distinguishing marks is the slightly reddish hue of the swamp sparrow's cap, while the song sparrow's cap is slightly more brownish. But it takes an expert to make this distinction out in the field. There are no adaptive explanations for the slight difference in color that distinguishes these species. Another North American example is the woodpecker-like common flicker (*Colaptes auratus*). This species occurs throughout the entire continent and is divisible into three color morphs, a yellow-shafted variety, a red-shafted variety, and a gilded variety. Each population is nearly identical in all anatomical characteristics except for distinguishing color variations on the underside of the wings. As with the trivial characters seen in sparrows, there is no adaptive advantage to having one or the other color variation. In fact the populations can interbreed, and offspring of hybrids show intermediate colors. Trivial traits such as these occur throughout the animal kingdom as well as in our own species. Can you say that there's any adaptive advantage to any of the slight differences you notice between people you see when you're out and about?

Natural selection could not have produced these traits, and competition cannot be invoked as any kind of mechanism in their formation. Trivial characters provide no advantage; they are selectively neutral, so they don't originate from competitive interactions, neither do they serve the individual in its daily activities. Throughout the second half of the twentieth century, more and more characters were found to be trivial. After DNA was discovered, much of it was found to have no selective value, suggesting a possible link between the ubiquity of trivial traits and their selectively neutral genes.[165] There was now accumulating a larger library of empirical data and theory suggesting that natural selection was not responsible for a great many biological phenomena (contrary to the neo-Darwinian "hard" selectionist view). This gave room to the consideration that competition might not be so crucial in evolution after all.

But the general public remained blissfully unaware of all these developments. While most people had heard of genes and DNA by the 1960s, it's safe to say that they didn't understand the implications of the intellectual history of heredity, or the simplifying assumptions (reduction of variables) in the quantitative models underlying evolutionary theory. One might assume that the wealth of biological data collected in the twentieth century, such as the discovery of DNA, or the sequencing of the human genome, or some other major development in life science, would have contributed a clearer understanding of competition at the dawn of the twenty-first century than the one put forward two hundred years before. Sadly this assumption doesn't hold. Except for a challenge from one emerging subfield of biology (see below), most people still cling to a forlorn belief that competition is a dominant feature of biology and human life. This might be due to the simplicity of its logic.

Competition's role in evolutionary biology today is nearly identical to what it was in 1900: The proof that competition exists is not so much a proof of observation, but rather a proof of reasoning. It was always possible to see evidence in nature of organisms fighting one another. Naturalists had been doing it for hundreds of years—watching dominant males fighting for dominance, for the right to mate with numerous females in a harem (as seen in the deer family, for instance); or the showy displays of bright feathers, elaborate songs, and dances by male birds trying to garner the attention of much duller female birds choosing the best mate. And there was obvious interspecies warfare too—think of the predator-prey relationships mentioned earlier. These hallmarks of biological competition have never been questioned. They serve the needs of a neo-Darwinian worldview perfectly. Natural selection favors slight variations that contribute to, and actually amplify, the adaptive traits that lead to more offspring, and better-developed traits. They may be adequate examples to satisfy the neo-Darwinians, but they aren't sufficient to answer whether competition is actually a primary cause for the entire evolutionary process.

That's why it must be conceded that competition serves as a proof of logic more than a proof of observation. The reasoning goes like this: Overproduction of eggs and embryos coupled with a lack of space and resources for all = struggle for existence (competition). This logic seems so sound that it's hard to see why anyone would want to criticize it. Indeed, it's probably the least written-about, fundamental principle in the field of evolutionary biology. Nonetheless, the emergence of a new view of evolution, such as one that emphasizes symbiosis over neo-Darwinian selection, puts competition squarely in the crosshairs. Such a science has been developing since the latter third of the twentieth century (see below).

This broad overview of evolution's intellectual history serves to show that one of the foundational concepts of the science, competition, is also one of the least scrutinized by biologists. It is an example of an underlying assumption that is so critical to the field that most people assume it must be something real, without clearly defining it. This has been called a "philosophical error" by one of the pioneers of modern biology, the late Lynn Margulis.[166] Specifically, she refers to competition as a "fallacy of misplaced concreteness," a common error in science. According to the philosopher Alfred North Whitehead, who coined the term, it is "an accidental error of mistaking the abstract for the concrete."[167] Competition is an abstraction. There are no units of competition to measure. It implies an agreement that opposing teams or individuals stick to in order to achieve some goal. For competition to make sense in modern biology there has to be clear-cut adaptive goals for every trait observable, and this clearly is not the case, as outlined above.

Today, natural selection and mutation are the standard explanation for evolutionary phenomena. Mutation is the "raw material," the ultimate source of variation in the organism, upon which natural selection operates to choose the most favorable genes. The hardening of this neo-Darwinian view, the creative application of it to human life (by particular scientists), and the rise of "popular science" in the mass media over the last forty years, created an awareness about genes and natural selection in the general public. The evolutionist Richard Dawkins had a profound effect on popularizing the neo-Darwinian worldview. His books, including *The Selfish Gene* (1976), *The Extended Phenotype* (1982), *The Blind Watchmaker* (1986), and *The God Delusion* (2006), have sold millions of copies. He has appeared countless times on national television in the United States and his native England, explaining the fundamentals of evolution, always with an emphasis on the gene as the target of natural selection. His coining of the term "meme," as a selfish replicating element of the mind, has become a colloquialism of pop culture, is now an entry in the *Oxford English Dictionary*, and is used ubiquitously in magazine articles, newspapers, and blogs to denote any idea that has "gone viral" on the Internet. As a popularizer of evolution, he has served a valuable role in raising an important science to the level of a public conversation. He has consistently defended science in the face of attacks from religious conservatives who would rather see evolution removed from school curricula. And yet his version of neo-Darwinism that has become popularized and accepted by the public—based on replicating elements (genes, memes, or individuals) acting in their own self-interest—perpetuates the misplaced importance of competition at the root of everything. Despite the fallacy of misplaced concreteness attributed to competition, the tradition begun by Weismann is little altered after 115 years, and has now crept into the consciousness of average citizens and intellectuals alike.

The biologist Lynn Margulis and her endosymbiotic theory ushered in a challenge to the popular view that everything is explainable as selfish genes and natural selection. She emphasized the ubiquity of symbiosis, and pointed to the fact that all familiar organisms are in fact amalgamations of genomes from different organisms. This implies a biosphere of interconnected populations and networks of organisms rather than one of constant warfare and self-interested individuals. What are the roles of competition and natural selection in this view? Clearly, the neo-Darwinian view of natural selection acting on individuals for the benefit of the genes they carry has to be deemphasized. In the alternative view of Margulis, natural selection is reframed as a measure of ecosystem efficiency (communities of interacting species). The biosphere itself is a selector, not of individuals, but of more efficient or less efficient communities.

This web of interdependence gives us a reason to make decisions that are good for the environment, and gives us a justification for taking care of our cohorts. In short it takes the emphasis off shortsighted selfish objectives that come from the belief in competition. In such a worldview we can restore competition to its rightful place—a defined contest for an agreed-upon goal rather than the driving force behind all human conflict, prosperity, and misfortune. But also we can see evolution in a more positive light, not as a brutal war of competing individuals but as a product of symbiotic relationships. The biosphere that we depend on also depends on us. This realization provides the impetus to act as caretakers, not only of other people, species, and ecosystems, but also of ourselves. In order to achieve any of this, however, we first have to come to terms with what we can and cannot control.

9

KNOW THYSELF, DON'T LIE TO THYSELF

If human life isn't a neo-Darwinian competitive process in which the most worthy prosper, then what are we left with to explain our station in life? The other common explanation for human behavior is free will. Even though it might not be immediately obvious, much of the belief in free will has to do with enemies, and evil, and bad things that seem to be everywhere around us. As mentioned at the outset of the Introduction, if we don't try to understand our enemies—or the sources of bad things in the world—we immediately assume that they are to be vanquished, and that they are different from us.

Most people don't question that evil is something real. That is to say, they believe it exists. Like an ethereal presence hovering over us, it's always threatening to settle into the dark crevices of our lives. The problem is that evil, hiding in out-of-the-way places, is not something empirical, but rather is identified as the works of people who decided to do bad things. Most citizens believe that criminals, terrorists, and enemy nations deserve punishment or elimination from society, since they exercised their free will by embracing evil rather than fighting against it, or helping others.

There is likewise an almost universal tendency among people—unless one is a philosopher—to believe that babies are born good, and with the God-given free will that they can exercise to escape the seductive pull of evil in the world. Ethicists and parents alike use this kind of reasoning to raise moral children and define appropriate social behavior. But what should we believe about inappropriate behavior? That it's also caused by free will?

In March 2014, a pregnant woman drove a minivan, with her three toddlers locked in as passengers, into the ocean at a beach in Florida.[168] As the mother drove into the heavy surf, she told her kids to close their eyes and go to sleep. Fortunately the minivan stalled out in four feet of water and the kids—none of whom could swim—were rescued by lifeguards and onlookers. The woman was arrested for child abuse, and the prosecutors began their investigations to decide if she had acted deliberately to kill herself and her children. Eventually investigators determined that her actions were premeditated and therefore deserved the charge of attempted murder in the first degree.

For her part the mother reported that she had endured fourteen years of torment by her husband and was determined to put an end to it the only way she could. She told her kids she was driving them into the surf in order to "keep them safe" from the abuses of their father. The husband denied ever harming his wife or kids, but believes that his wife suffered from extreme mental illness. If that was the case, was the woman acting of her own free will or not?

Our society's sense of justice is based on the idea that we can be judged by our actions. If someone is mentally competent, and not coerced or held hostage, and they decide to do evil, then he or she deserves harsh punishment. This is almost never questioned. But I *do* question it. Whether this mother was sane or mentally ill (a doctor or a judge can decide this), her judgment about what was best for her children shows low fitness if it puts her offspring at risk. We have to assume that her social and physical environment acted to bring out her actions even if the stimulus was a developmental brain defect or lesion. The state of her brain was affected by physical changes from previous experiences, and this too had a role in her behavior. Both environment and development (mediated in part by genes, and in part by developmental experiences) should be the main considerations in judging her actions. This approach contradicts the idea that she was actively engaged in exercising her free will.[169]

I'm not concerned whether the mother in this case is guilty or innocent; I'm not a legal scholar, nor particularly interested in the case as a human-interest story. But I am deeply concerned with how this incident, and so many other stories that we read about or experience in our own lives, fits in with the unquestioned social narrative of free will. This woman was no more free, in my view, than any of us are when we make choices of a far more benign nature in our day-to-day dealings.

I'm a very affectionate person. I give my children and other family members hugs countless times a day. Even if I just hugged them when I walked into the room, I hug them again when I walk back out, even if it's only after a brief chat of a couple of minutes. I could intellectualize this behavior and say it's obvious that I am trying to encourage them in all their activities through embraces and reassurance, or any number of other rationalizations, but this is not a conscious thing. My affection is automatic and governed by neural processes over which I have little control—neural processes that release hormones for affectionate responses. Sure, I could curtail the actual behavior, perhaps, if it made them uncomfortable. But then another neural circuit would be activated—and with it a whole suite of hormones related to protection of loved ones would kick in. In my worldview nearly all behaviors, not only simple ones like hugging and showing affection, or protection, are determined by large suites of stimuli, most of which we are blind to as they occur.

We assume we act with deliberation, but in actuality we are just highly complicated automata reacting to hundreds or thousands of simultaneous stimuli. Social standards and cultural norms help us to process the stimuli. They train us to act predictably, and to respond uniformly to the multifarious activities that go on in our daily experiences. Think of something like a handshake. A handshake is now an almost universally acceptable form of greeting. If you choose not to shake someone's hand, because you foolishly think you are exercising your free will, you will send a strong message to everyone in the room. But it wasn't free will when you acted so rudely; instead it was some other stimulus—perhaps from another meeting, perhaps from a dream lodged in your long-term memory, perhaps some lingering animosity from an angry thought that was so fleeting you don't even remember having it—that made you keep your hand by your side and insult the person whose hand you rebuffed.

Making sense of complicated behaviors, even something as seemingly simple as a handshake, comes down to a study of neurobiology, the study of cells that make up the brain and other nervous system components. Some things to consider: There are more connections between neurons in a single human brain than there are stars in the Milky Way. In fact, there are more than 100 trillion such connections.[170] How those connections become linked is due to genes and the environment of the developing child. When a child is born it has an even greater number of neurons than it actually will use in its lifetime. Which neurons persist and which ones become deactivated is a process of natural selection. The population of neural cells at the adult stage and their interactions is a selected subset of the original population at the infant stage. Selection happens through the experiences of the developing child.

One of the iconic examples of natural selection in nature happened in England during the Industrial Revolution. In the forests around England's larger cities lives a moth, the peppered moth (*Biston betularia*) that settles on tree trunks during the day for rest and basking in the sun. The color of this moth, originally adapted to light-colored lichens and light bark coloration, was light cream with small dark blotches of black pigment (called melanin). As the cities grew larger, factories and fireplaces produced chimney smoke, full of particulate matter (soot), that darkened the trees, killed the lichens, and created a landscape of nearly black tree trunks. In such an environment the light-colored moths were "sitting ducks." Their high contrast with the black sooty substrates made them easy prey for birds that hunted during the day. The ones that were best able to escape predation were moths that contained more melanin. Darker-pigmented moths had a better chance of not being detected and were better able to reproduce and pass on their dark coloration to offspring. Eventually the population lost nearly all of its light-cream-colored moths, and today, near the industrial centers of England, the dark form of peppered moth is most common. The population was naturally selected for its darker coloration, and this resulted in adaptation to the new environment of black substrates caused by the Industrial Revolution. This adaptation is often referred to as "industrial melanism."

Just like a species of moth that becomes adapted to dark substrates due to industrial melanism caused by its environment, the cultural and social milieu determines which neuronal connections will be most important for a developing child. In nature we call this natural selection, a phenomenon that is measured over generations of reproducing individuals. But in the brain it's called neuronal selection, and it is measured over stages of brain development from embryo to adult. Put simply, a baby's brain is a different organ than is an adult brain. The population of neurons that you were born with made numerous connections and networked together in complicated ways that were significantly influenced by your cultural and social conditions. If you had been placed in a different social setting, or a different culture, your adult brain would look different to a microbiologist, and it would consist of vastly different connections than it now does. In electrical terminology, we would say that the circuitry had been wired differently.[171]

Just because we had no control over how our brains were wired, however, doesn't allow us to claim innocence when we behave badly. It's true that we had no control over the neuronal adaptation of our upbringing, and that constitutes grounds for claiming that free will is nonexistent. But the lack of free will doesn't mean we have freedom to do harm. For our purposes, it simply demonstrates the shortcoming of using free will as an explanation for our station in life. How we came to garner success or wallow in failure is better explained as a population war: a neuronal adaptation to a huge suite of stimuli throughout life that acted as a selecting mechanism for how our brains are organized.

Neuroscientists have determined that the human brain, particularly the cortex, or outer layer—the area that is responsible for reasoning and planning[172]—is densely packed with neurons, more so than any other species. Humans, in fact, have far more neurons relative to body size than does any other organism.[173] This means that one of our most important populations of cells exists between our ears, as brain tissue. This population increased to its current size within the first few years after birth. Like all populations, this one too was subject to natural selection, and its individuals grew and formed connections to one another based on their state of activation during experiences in the life of the host organism. These experiences ranged from eating and talking to more traumatic ones like falling and fighting. How we process each moment now is heavily biased by those experiences. In fact, how we perceive momentary stimuli, our "joys and sorrows, memories, and ambitions, sense of identity and free will, are no more than the behavior of a vast assembly of nerve cells and their associated molecules." This is what the Nobel laureate Francis Crick (who with James Watson and Maurice Wakins—and their colleague Rosalind Franklin—discovered the structure of DNA) has called the "astonishing hypothesis."[174] All the stimuli that we are experiencing at this very moment are filtered through a population of cells that may have been selected for vastly different stimuli. Therefore our decisions and actions cannot be considered free from these biases.

If we accept the findings of neuroscience, we have to conclude that our behaviors are due to these neurons (nerve cells) and their connections in the brain. Far more connections are produced (up to ten thousand per cell) than are actually needed to sense and process external and motor stimuli (the impulses that give rise to our actions). As a youngster develops, the experiences she encounters have an effect on these connections. The redundant ones and the ones that do not participate in any of her experiences are eliminated. The ones that participate and aid in her success in navigating her environment persist and therefore are "selected."[175]

I would suggest, therefore, that there is not only very little room to make a materialistically based argument for free will, but that it's actually worth questioning the very existence of anything like free will. What you believe about your purpose and direction in life is governed more by the circumstances and history of your population than it is by your own will to act freely. I recognize that this is a universally unpopular idea. Nearly everyone across the intellectual spectrum—from the most liberal philosopher to the most rigidly conservative theologian—is wedded to the idea that free will exists. Most people seem to want to believe that humans are unique and autonomous, that we determine our own fates, and that our good choices and the resulting happy outcomes are the result of our superior abilities (proved in competition) and good choices (free will). The other possibility—that the rough trajectories of our lives have already been set in motion before we are born, that some of us are dealt good hands and some bad—is at odds with most people's cherished beliefs.

The roots of free will as a worldview lie in theology. Thomas Aquinas believed that free will was a gift from God to his favorite creation (humankind), and this idea is at the foundation of Christianity. All the variations of Christian beliefs say that you as a person can either choose to accept God's gift, exercise your free will, and do good—or you can reject his gift, choose to do evil, and go to hell. That decision—to pick sin over salvation—is nobody's fault but your own. Therefore you deserve punishment if you've done something evil. By this reasoning, however, somehow the rational belief that killing another human is never justified becomes used to rationalize the death penalty.

Of course this reasoning isn't new. The Catholic Church had the authority to dole out such punishment in historical times. Religious punishment was often pushed to horrifying and clearly illogical extremes. In the Tudor era British subjects' only real religious obligation was to worship as the king or queen instructed them to. It was understood that so long as they obeyed the monarch's decrees about religion, they had fulfilled their obligation to God. However, in the fifteenth and sixteenth centuries Europeans on the Continent, most famously under Martin Luther of Germany, began to resist this idea and demand the right to worship as they saw fit. These people were judged to be heretics and punished by being tied to the stake and burned alive. In England, as King Henry VIII and subsequently his daughters, Mary I and Queen Elizabeth I, vacillated between the old Roman church and the new Protestant church, Protestants and Catholics—one supposedly "good" and the other supposedly "bad"—were sometimes burned on the same day.[176]

Free will and punishment have always been bedfellows, but the concepts have also made for an easy moral narrative. The concept of free will has become secularized in the modern world. Even if you are not religious you have free will, according to the legal system. Laws in our country guarantee "freedom of religion," which implies that you don't have to buy into Christian theology. Therefore, to follow U.S. laws, you should be able to ignore the concept of free will as a gift from God. But federal law also defines murder in the first degree as premeditated, and undertaking the act is a felony because it was done of one's "own free will." Therefore free will is one fundamental tenet from theology that transcends religious belief and to some degree forms the backbone of American civilization. When you ask many secularists, Why do you think we have free will? they'll often reply by saying, What else is there? It's a cop-out. There is a vacancy here, and it is is something that philosophers are still struggling to explain fully. The important idea—and one that is generally ignored today—is that there are historical circumstances that have nothing to do with free will but have nonetheless brought you to the place you are at today and continue to influence your actions. These circumstances can't absolve you of wrongdoing, but they can certainly clarify the intention behind it, and possibly gauge the appropriate punishment.

I believe that the idea of free will is pleasing to most because it allows us to take ownership of our successes and neatly to blame others' failures on poor personal choices. Perhaps most germane to our discussion throughout, the idea of free will allows us to justify our desire to vanquish other people. The existence of "evil" is manifest in people who do evil things, and they are therefore our enemies. It's very easy to transform blame for individual misdeeds, by criminals or terrorists for instance, to an entire population. Human warfare, in this instance, stems from an idea grounded in theology that has no basis in materialistic science or modern understanding of human behavior (that is, data from neuroscience).

Your political and cultural tastes were mostly likely set in motion by your parents or the social milieu in which you grew up. You may have chosen to react against your upbringing, but that reaction is no more free will than falling in line with it. Just as with the idea that "life is a competition," what we perceive as "free will" can be boiled down to the common human foibles of "self-importance" and hubris. When we give historical circumstance its fair due, we get a clearer picture of how we came to our present situations. So, if our lives aren't merely guided by competition, or by free will, or by some benevolent deity, what guides us through the maze of life? Populations profoundly drive the path we take: those that compose us, those to which we belong, and those with which we come into contact.

I recognize that my worldview with respect to free will may be difficult for most people to digest. Many of those who have written on the topic of free will have been philosophers, and they have spent considerable amounts of time persuading their readers that the topic requires wide reading in philosophy, ethics, theology, or some other intellectual subdiscipline. But I don't think the topic of free will has to be so complicated. The unlikelihood of free will can be understood from the basic facts of evolutionary theory. If we are all related by descent, then our traits came from our ancestors. Of all the conceivable common ancestors that link humans with great apes, and those ancestors to primitive mammals, and those ancestors to quadrupeds (four-legged animals like reptiles), and those ancestors to something fishlike, where along the way did anything akin to free will evolve as a trait? Consciousness is not the answer. There are plenty of living species that show varying degrees of consciousness, but no one includes them in the discussion of free will. If it's just varying degrees of freedom of movement and choice, then we can definitely see a gradation of those traits in living animals. Terrestrial lizards, for instance, are restricted to warmer habitats (being "cold-blooded") than are mammals. Does this mean that mammals have more "freedom" to choose colder habitats? Mammals have fur as insulation, which means they can be active over a wider range of temperatures than reptiles can. But this isn't free will, it's evolution of physiological adaptations that may have brought with it changes in brain anatomy. Mammals tend to have larger brains than other vertebrates (relative to their body size), which is a rough guide to their abilities in "decision making." (It comes as no great surprise to anyone who has ever had pets that dogs make a lot more seemingly autonomous decisions than pet lizards do.)

The problem is, the dog's decision making isn't really free, it's based on stimuli—the jangle of keys means "Let's go for a ride," or the shaking of the box of Milk-Bones evokes spasms of excitement, or some hormonal cue from a full bladder makes it pace in front of the patio door because it's time to "do her business in the backyard." Dogs express varying degrees of emotions, but no one would argue that they are free to determine their decisions and long-term goals in life. Lizards also react to stimuli, but they are more basic—often temperature related or visual. Reptilian pets often are immobile in their cages, moving only in response to particular feeding routines. Fish are less free still. They are either frantically swimming, trying to find a way out of their glass-walled predicament, or totally still with only occasional gill contractions, until tiny flakes are added to the water's surface, at which time they go for the closest ones. Where along the spectrum of behaviors in living vertebrates can we perceive the rudiments of free will? More relevant, where in the course of human ancestry, among our ape cousins, perhaps, could something so profound as free will evolve? Even the most intelligent apes are not thought by philosophers to have free will.

The problem, as most philosophers see it, lies in the realm of something called "determinism." This is not something many biologists discuss because behavior in organisms, including humans, is considered to be a huge suite of traits that can only be measured rather coarsely. The philosopher insists, however, that deterministic laws should be present in organisms, like the deterministic, predictable laws that affect gases and molecules in chemistry and physics. They say that human behavior, if it is in fact governed by cause-and-effect determinism, should show much more predictability than it does. We don't, as they like to point out, act or feel like programmable robots, so how can someone suggest that we aren't free? Sidestepping the rapidly accumulating data in neuroscience (which does in fact show plenty of cause-and-effect activity in the human brain), free-will advocates claim that humans have the ability to break free from any stimulus, and choose their own course of action.

The entire discussion of free will, then, has conventionally centered on determinism, much to the philosophers' delight.[177] On the one hand, those who claim that free will exists are weaseling around the issue of determinism by ignoring biological causes and their effects on organisms. On the other, those who claim that free will doesn't exist have the burden of explaining how humans, with their seemingly limitless array of variable and unpredictable behavior, represent determined entities at all. For the biologist, however, the discussion seems moot. Philosophical determinism makes sense as a mind exercise, but it doesn't help us understand the evolution of our species.

Every single trait exhibited by humans is the product of genes, embryonic development, and the environment. If free will is something real, it too has to be an observable trait that has developed over geological time, in our ancestors, and must be subject to evolutionary transformation. A suitable conclusion for some zoologists (those who choose to engage in debates about determinism) might be that human ancestors showed varying degrees of consciousness, and that such a discussion then puts great emphasis on the brain as the organ that allows humans to have free will. Some tetrapod ancestor (such as a mammal-like reptile, for instance) clearly had no consciousness—we have fossils that show, for instance, that early mammals didn't have a well-enough-developed cortex to qualify as a conscious being. We can compare fossil brain cavities with those of living species and recognize a spectrum of probable mental capabilities in mammalian ancestors. In general we can say that primates have the anatomical features that indicate superior attention abilities, and that this highly adaptive trait has been amplified through human evolution. Whether or not this led to consciousness is debatable. It is, for instance, possible that consciousness was simply a by-product of traits that evolved for greater attention and focus.[178] Nonetheless, a discussion of consciousness may be a satisfying way to understand the evolution of the brain, but it is not what free will advocates have in mind. To them, free will is the ability—despite the constraints of genes, embryonic development, and environment—to overcome the determining factors of human behavior. It is, in simplest terms, belief in the miraculous.

Many scientists I have polled believe that organisms are the products of genes and their environments, and that's it.[179] These same scientists, however, are reluctant to admit publicly that there is no free will—not because they think it exists, but because they want to avoid controversy. Many more, I'm sure, simply don't want to go on record. There is great tranquility in the sea of science so long as controversy remains stranded on the shore. This reluctance to discuss free will as an evolvable trait is, to me, the best indication that the topic is simply taboo. Too many people, scientists and nonscientists alike, are too disturbed by the prospect of no free will ever to give it a fair hearing.

I understand the sensitivity that people feel when they hear philosophers discuss free will and determinism. When only those two extremes are put forth—either you're completely controlled by causes or you're free to do what you want—who wouldn't choose free will? That's what disappoints me so much about the issue. It's more productive to shift the conversation away from determinism, and accept that we are in fact the product of causes—historical and circumstantial—but aren't predestined for some "determined" outcome in life. Through an awareness of causes (that is, education) of our actions and positions in life we can actually make informed choices to some degree, if we care to, and change the course of our lives. It is this hopeful view, based on the rejection of the idea of free will, that can help us become better stewards of ourselves and—if this viewpoint is amplified through teaching—of the entire population.

There is, furthermore, another population effect that rejection of free will might accomplish. Doing so would bring with it the evaporation of the mythos of evil. The moment you stop using free will as a justification for punishment, your understanding of evil changes. No longer are other groups judged in relation to absolutist ideas of right and wrong. Perhaps most important, if you reject free will you are instead forced to focus your attention on the historical and immediate circumstances of populations, and those are empirical facts that can be collected and treated scientifically. What is right for one population in time and space might not apply to another. The judgment of right and wrong is relative to the population in question. When we reach this stage of intellectual progress, bigotry and prejudice will be things of the past, and we shall begin the larger task of global stewardship.

Ultimately, taking care of others is the windfall that comes from discarding the idea of free will. But this stems from understanding ourselves, as Socrates himself implored us to do. As we have seen, the self is riddled with and connected to populations of others, and is therefore far more complicated than the ancient philosophers ever could have imagined. Self-betterment is really about monitoring and caring for the populations to whom you're connected.

At some point in my life I realized that I was never really happy if those around me in my professional and personal relationships were unhappy. Furthermore, the goals we set out to accomplish together, whether it was a family gathering, hanging out with friends, undertaking a concert tour, or cowriting an album, were hampered if any one of us wasn't feeling well. So eventually (I can't say when, exactly) I began to feel a sense of responsibility to the relationships, and recognized that I had to take care of myself, so that I might be as good a participant as possible in them. It feels instinctive in some ways to be selfish, but what I recognized is that selfishness never feels good by itself. Taking advantage of others is shortsighted and lonely. But taking care of your own needs so that you can function better in relationships is key to all human social endeavors. By extension, this realization—that the selfishly motivated desire to feel good is enhanced by the benevolence and generosity we give to our families and cohorts—can be applied to the other populations with whom we coexist. Indeed, such an objective could serve as a humanistic pursuit for our future.

Earlier in this book I mentioned that 99.99 percent of the species that have lived on our planet have gone extinct. This is alarming to many who are uninitiated into the realities of biodiversity. It seems logical, though, to consider the eventual fate of our own species in this regard. We want to make sure that we remain in the 0.01 percent who don't go extinct.

From a purely naturalist/evolutionary perspective, human beings are just another species, a biological "accident" that slowly evolved from apelike ancestors. From this perspective we can assume that eventually we are going to end up as most every one of our ancestors: extinct. Bear in mind that the fossil record shows that most large mammalian species, such as our own, live only for roughly 2 million years before they die out. If you consider the two-million-year-old fossil *Homo habilis* a "human" rather than an "ape," then humans are already very old. Perhaps we are precariously close to extinction simply by the law of averages.

The naturalist in me believes it's foolish to look at humans selfishly, as something special in the grand scheme of evolution. The humanist in me, however, sees something else. A humanist perspective acknowledges the uniqueness of *Homo sapiens*, recognizing unique abilities we have in relation to all the other species with whom we share the planet. We are the first species to attain true consciousness, to recognize and ponder our own existence, and to theorize about the future and our eventual fate. So maybe our goal should be to embrace the fact that we are the exception rather than the rule, and thus focus on our species' longevity. This requires a shift from merely short-term predictions about social, economic, or political trends to predictions and planning for populations within the framework of evolutionary time. We need to think in terms of thousands of generations rather than one or two generations of human lifetimes.

This shift in focus is a foundation for an environmental ethic—a long-term sustainability protocol that benefits our entire species, not simply a particular political, national, or cultural group. And what about the other species on the planet? I see them as players in the long-term sustainability in the same sense that a landscape gardener or horticulturalist views the inhabitants of his plot, full of ornamental flowers, shrubs, maybe some food plants, and cultivars—and many that aren't apparent to the naked eye, such as microbes and fungi that subsist in the soil.

The successful stewardship of the biosphere—the home of all the raw materials that sustain us—requires everyone to be educated and mindful about their own impact on it. The answers to basic questions should be common knowledge, such as: Where does freshwater come from? What happens when we burn things? How do motors work? What is energy efficiency? What is the ultimate source of the food that we consume? No single individual can be an expert at answering all of these, but human curiosity can be trained to look in such directions. We could insist that our kids get a basic natural-science education. Eventually a culture of sustainability could become ingrained, just as children who are raised around gardens develop a reverence for plants simply because they grew up in a social climate that valued cultivation.

A lot of environmentalists seem misguided in their quest. In some cases it appears as though they care more for other species than their own. I would consider this overly sentimental attitude a weak kind of environmentalism. It might achieve an admirable goal (perhaps to save an endangered species), but it is underlain by a weak intellectual foundation:

Preserve everything. This might be a holdover attitude from a strong theological tradition, as if all species originated by God's creation; and thus all species are equal. Or it might legitimately be a knee-jerk response to the carelessness and reckless degradation of fragile habitats by industrialists in their quest to convert natural resources into profits. In some ways I can sympathize with such views; I too dislike machinery in areas that are designated "forever wild." But in general I disagree with environmental attitudes that are too emotionally charged or willfully ignorant. If we are going to achieve a lasting environmental ethic, it needs to be divorced from such sappy sentimentalism. We need to act like stewards and land-use managers. Sometimes certain species need to be favored because they are more important to the ecosystem. Our major problem at the present time, however, is the loss of habitats and species due to human industrial practices, at a rate too fast to let us discover which are the most important to us. Until we have more robust ecological data, we have to assume that *all* species are important in the current equilibrium of coexistence. We can't reestablish an ecosystem after the extinction of its species.

Part of human exceptionalism—the belief that we are entitled to be stewards of the biosphere—lies in the fact that no other species has had the physiological or neurological capacity to recognize and plan for its own continued existence. In short, this correlates to the possession of consciousness. Our attainment of consciousness entitles us, but it also burdens us. I think our ability fully to understand our surrounding world and its importance to both our productivity (industry) and longevity is what humanism is all about.

If we decide that our species is unique among all species and that we deserve to be the top priority, then our goal is to develop an ethic of preservation that benefits our species' longevity in geologic time. Once we reach such a decision we can talk about "right" versus "wrong." If we know, for instance, that pollution causes irreparable damage to the biosphere, then it is wrong to pollute because every human depends on the biosphere. We won't persist long if our ecosystems are drastically depleted of key species that depend on healthy conditions. Pollution may cause extinction, and therefore it is bad.

Conversely, if we decide that humans are no more important than the other species on the planet, then there is no criterion for maintaining any kind of ecological balance. If we try to preserve everything, for instance, then the invasive species, and the parasitic species, and those that have benefited from human-caused extinctions of native species—species already "out of balance"—will take over from our lack of stewardship. Humans have already altered the biosphere in nearly every important realm, from every freshwater habitat to virtually all the arable crop land, from the global nitrogen budget to the balance of CO_2 in the atmosphere.[180]

Perhaps the greatest challenge we face in the twenty-first century is managing the evolution of other species. This idea is gaining steam and falls under the heading of "evolution management."[181] The debates will rage, of course, about how to tweak the biosphere in order to attain or preserve a "healthy" ecological equilibrium, while still acknowledging and practicing the traditions and lifestyles of a modern industrialized species. We may have to give up on certain technologies in the interest of sustainability. In short, this is precisely the kind of issue that we face as stewards of the planet.

One obvious attempt at evolution management is to slow the rate of increase in the human population. Unfortunately, if we pursue this idea to its logical conclusion we have to decide on policies that trouble me deeply and are questionable from a humanist perspective. For instance, the ideas that people shouldn't be allowed to have children, or that parenthood should be allocated by a lottery system. To me these ideas are unacceptable and as ethically unpalatable as any policy that justifies pollution. The only lasting change comes not from harsh restrictions, but sensible policies that resonate with the bulk of the population. At present most of the world believes that people have a "God-given" right to procreate, just as many landowners or mining companies believe they have a "God-given" right to pollute their land if they want to. But exercising these rights is not good for the population.

Some noted scientists believe that there is no longer anything we can do to avert climate disaster. James Lovelock, a noted English chemist, medical scientist, and inventor, proposed that the biosphere and the inorganic (nonliving) parts of Earth are intricately intertwined as a complex interacting system. His Gaia hypothesis suggests that the biosphere itself can be considered an organism that is self-sufficient and self-regulating with respect to the nonliving parts of the system. According to Lovelock, we have poisoned this organism to such a degree with our toxic emissions (primarily CO_2 in the atmosphere) that we are beyond the point of self-regulation, and now the inorganic parts of the system (reflected in rising global temperatures) will make most of the biosphere extinct and most of Earth uninhabitable.[182] Many other environmentalists believe that we would have had to dramatically curtail our carbon emissions beginning in the early 1960s if we wanted to maintain the health and self-regulating properties of the biosphere.

While I appreciate Lovelock's creative scientific career, and I consider myself allied with environmentalists in spirit, I differ from this dark outlook. I believe there is still some hope. We are the only species fully aware of our own circumstances on a global level, and as a result we have the chance to alter them. I don't think we should take any pride in this; after all, we didn't will ourselves to this position through any effort of our own. In fact there is a burden of responsibility, as I mentioned earlier, that comes with our awareness; we are the first species to recognize the importance of preserving our environment—both for the sake of other species and for our own. How we choose to handle that burden will inevitably impact other species. Some of them may become cherished or idolized as symbols of a healthy environment, for the benefits they bring to mankind in their ecosystem services. Others may not survive.

One reason for my optimism stems from my belief in mass media and its power to change human behavior. I witnessed a human experiment within my own lifetime, although it was never designed as one. It was simply a commercial that ran on national television for a couple of years during a campaign by the "ecology movement." Only forty years ago, when I was a child, the sides of the interstate highways in this country were littered with trash. Most citizens of the United States didn't think twice before throwing a half-eaten sandwich, or dirty diapers, or used Kleenex, or even a bag of household refuse out the window of their speeding car. The average citizen at that time was simply following a protocol set by nearly every factory and citizen in the country since the nineteenth century.

Polluting streams with industrial waste was also the norm in those days. Nearly every factory in my hometown of Racine, Wisconsin—like those in Milwaukee, Chicago, Gary, Detroit, Buffalo, and other cities of our industrial heartland—were built along major streams and rivers. Why? Well, one reason was certainly for transport. In the nineteenth century nearly all goods, raw and finished, were delivered by boat. But there was another convenience in those days before anyone had any awareness of ecological health or environmental impact. The rivers served as massive toilets for the industrial chemicals that were by-products of manufacturing. Engineers knew nothing about the toxic effects of their industrial chemicals on the environment, but they knew that many chemicals were toxic to humans. So what better way to get rid of them than having a constantly flowing river right outside to flush away the poison, and send it continually downstream?

Millions of gallons of chemical waste were haphazardly spilled into the rivers of American cities for generations, poisoning fish, fouling the freshwater, and collecting in the sediment. Some of these areas are still toxic today, even though the cleanup began in the 1960s. It's true that many cities, such as Pittsburgh, were heavily polluted, but in forty or so years they have made incredible advances in the rectification of their toxic past. Some chemicals can be neutralized, but others, such as mercury and PCBs (known carcinogens), may never be made inert, and they continue to be concentrated in the riverine ecosystems throughout the nation.

By the time I came along in the 1960s, awareness was growing, particularly about PCBs in rivers. But the average citizen wasn't educated about pollution at all. Most people just assumed that if you threw a bag out of your window it would break down by weathering processes eventually. This type of reasoning made the entire interstate system a waste dump. From my earliest memories, traveling between Wisconsin and Indiana, I can't say I remember littering, but do I remember taking notice when the "ecology movement" began.

I can't even say which organization began getting us kids involved.[183] I was in fourth grade before there was even a national law passed for safe drinking water. But around that time I remember getting certain cereal boxes specifically because they had "toy surprises" inside. Kellogg's put brightly colored green stickers and magnets inside Raisin Bran. My brother and I collected all the different varieties. These "ecologies," in green with white backgrounds, were etched into plastic as nifty-looking logos. These symbols of the environmental movement really got me and my brother talking about ecology and what it meant.

As part of the "ecology movement," a memorable TV commercial was shown repeatedly throughout the week during prime time. An Indian chief, sitting proud and majestic on his horse, is at the side of the highway with America's "purple mountain majesties" in the background. Suddenly a station wagon full of lighthearted vacationers speeds by, and a bag of trash is ejected from their window and splatters at the feet of the chief's horse. The TV audience is left with a closeup of the Indian's face. A tear wells up and rolls down his cheek as the screen fades out.

This campaign, called "Keep America Beautiful," got people thinking. Not only did the interstates quickly get cleaner—community groups were out in force to clean up the roadside—but people stopped littering. Today the interstates are remarkably clean. Millions of cars pass by the busiest interchanges and not a single scrap of waste can be found along the medians or grassy shoulders. It is only the very rare careless person who litters now. Almost every citizen knows that litter ends up in the rivers, reservoirs, and drinking water. Almost every citizen knows that toxic chemicals have to be properly disposed of. The public has indeed been educated, and this has led to habits that we take for granted. Holding on to your trash is a habit, not an instinctive behavior. Nearly everyone shares the habit of having a trash bag in the car and waiting until the next filling station to throw it away properly. Some, like my own family, separate trash into recyclable stuff, compostable stuff, and stuff for the landfill. These are the kinds of learned behaviors that become habits and deeply impact the environment.

In one sense this example illustrates how your will to choose between two alternatives—self-preservation or careless destruction—can be affected by education through mass-media campaigns. In the quest to become excellent stewards of our planet, this is a bright spot, in my opinion. But in the bigger picture it is less a question of choosing between two alternatives than it is a reason to work toward mass education across the population. Creating empathy and awareness affects what choices are made across an entire population. Instead of debating whether this is free will or not, let's just recognize that a historical mass-media experiment forty years ago led to a massive change in people's "freedom," and no one is complaining. I never hear someone lamenting the fact that he doesn't feel free in this country because he can't toss trash out his window. Everyone sees the upside to the broken habits of old, and none of us were entirely free in the experiment.

I guess the impact of the "ecology movement" in the 1970s never left me. After all, I was drawn to biology in high school and went on to get a Ph.D. in zoology. Furthermore, I took the concept of mass-media's influence to heart and made a conscious decision (at least it felt conscious, but more likely it was instilled in my youth by that pollution commercial) to use it in my professional life. I have an unusual platform, where music fans sing along with me at concerts I perform. They're singing along because they know the words by heart. The songs I'm singing, however, are not love songs. In my band we write songs about current events, the environment, war, religion, and science. By using music that sounds cool I hope to affect listeners the same way that image of the Indian on the side of the highway affected television audiences of all ages and creeds. I'm very fortunate. Bad Religion is very popular the world over. Our songs reach a lot of people. I hope I will be forgiven for letting out this secret to my songwriting and singing. But I use the opportunity and privilege of being listened to to try and persuade the audience to be concerned citizens.

I believe that it's time to take an active approach and try to be self-proclaimed stewards of the environment. We will have to accept that many species on the edge of extinction might die out. We will have to reconcile ourselves to compromise with big business and large-scale polluters, and accept that the jobs they bring or the energy they create are sometimes worth the trade-off of regional environmental loss. Hopefully, however, as green industries become the norm, pollution will abate. Individuals will help themselves by creating a balanced life and monitoring what they consume and throw away. Across the population such individual stewardship will be manifested in a more benign coexistence with other species.

We like to flatter ourselves that we are the most exalted of all sentient beings in the biosphere. While we are special and unique in many ways, we are not immune to the hidden hand that guides population wars. The myth of free will has provided us with a comfortable lie: that we should ignore the environmental influences and historical circumstances that shaped us. The fallout from this ignorance is that there is no generally agreed-upon understanding of human nature, and therefore we cannot hope to produce an agreeable agenda about what's best for the maintenance of our species. In order to be stewards of the biosphere, natural history must be placed at the forefront of a scientific education. Part of this agenda requires a strong emphasis on learning and teaching about the "self," not as a stand-alone, free entity, but rather as a nexus of activity for other organisms that are crucially important for its happiness and survival.

10

EVOLUTION MANAGEMENT

In the time frame of evolution, the ultimate fate of all populations is extinction. Extinction is the moment when the last surviving individual of a population bites the dust. Some populations have gone extinct in little more than a single human lifetime. The dodo of Mauritius was so unwary that early visitors to the island, in the 1600s, found the flightless bird to be easy prey for food. However, it is likely that hunting was not the sole cause of the dodo's extinction. Remote island populations (such as those on Mauritius) are often much more vulnerable to population fluctuations, particularly in the face of habitat destruction due to human activities and colonization. Mauritius's forests (the dodo's preferred habitat) were being cleared for timber precisely at the time the dodo population was declining. By 1700, a hundred years after it was discovered, the dodo was extinct, along with several other large vertebrate populations on Mauritius.

Populations are interconnected in complex ways that are often hard to unravel. The dodo's extinction coincided with the near extinction of a tree, *Sideroxylon grandiflorum*, known as the tambalacoque, or dodo tree. Some biologists have speculated that the trees' seeds needed to be processed by the dodos' gravel-filled gizzards before they could germinate. Others have suggested that the real culprits were the domesticated pigs and goats who trampled the soil and grazed it bare, making germination of the tambalacoque impossible. These animals invaded Mauritius in swarms along with the first Westerners, and widespread deforestation followed in their wake. Whatever the case, the fate of the dodos, trees, and the invasive species were linked; the original species, vulnerable from the start, declined as the newcomers thrived.

Extinction is caused by unobvious, often multifarious factors. It's not always clear when a species has gone extinct. Other species are artificially sustained by human intervention. The tambalacoque is prized for its hardwood; modern Mauritian farmers now help the seeds germinate by abrading them, or running them through gem polishers. It is still unclear if this action is crucial to the species' survival, but either way we can be certain that the tambalacoque lives on at least partly because human beings find it too valuable to be allowed to die out.[184]

Many American families have witnessed a real-time population crisis in a native tree species. If you live east of the Mississippi you've probably heard about the demise of elm and chestnut trees. Most small towns in the eastern United States lined their main streets with elms. The street signs still stand (how many of you readers live on an Elm Street?), but the trees are long since dead. Our grandparents may remember these enormous elms shading their streets in summertime. But by 1960 tens of millions of mature trees had died and were cut down. Their death was due to a beetle infestation from Europe. Logs from Holland, destined for furniture makers in Ohio, brought the beetles to our shores. And the beetles carried a fungus of the genus *Ophiostoma*, which infests and kills mature elms.

Another arboreal treasure, the American chestnut, met a similar fate. These graceful giants were loved as shade trees, nut producers, and beautifully ornamental and fragrant springtime bloomers. Their tall trunks made lumber that was long, straight, workable, and strong. Many of the old barns that still stand throughout the United States are held up by chestnut beams fourteen inches or more in width. In fact, the old barns that dot the countryside of upstate New York couldn't have been built without the support of these beams. The size of these barns was limited only by the length of the single beam that supports the roof, and their huge size is a living testimony to the majesty of those old timber stands. Despite its usefulness and healthy population size, the chestnut declined in the twentieth century due to a blight (fungal infection) from Asia. Imported trees carried a fungus from China to America in the first decade of the twentieth century, and it quickly spread to nearly all the chestnuts on this continent. Aggressive logging in the 1920s (to prevent the spread of the blight), along with the disease itself, led to the destruction of most of the timber. The population was nearly decimated; it went from an estimated 3 billion trees at the start of the century to fewer than one thousand mature trees today.

Remarkably, however, the chestnut persists against the odds; some small stands still exist, one of which was planted by an early settler in Wisconsin in the 1800s. These trees are the lucky remnants of a once larger, healthy population. Somehow the remaining stands had a fortunate combination of genes and habitat selection that led to lower susceptibility to fungal infection. One wonders how much of this potentially resistant variation was eliminated by overzealous logging of chestnuts to quell the spread of infection in the last century.

If you look carefully, you may yet find elm trees in the forests of the eastern United States. They aren't large, they aren't dominant in number, and they aren't creating shade the way their ancestors did, but they are in fact growing and reproducing. As with the chestnut trees that you also may find growing in out-of-the-way groves at much-reduced number, these individuals show some degree of resistance to the infestation early in their growth. The problem is, when the elms reach a certain age, usually much younger than their ancestors' age at maturity, they succumb to the elm beetle and the fungus it brings.

The key to this lesson on trees is that the populations were not vanquished even though all indicators suggested that they should have been. Unlike the dodo, a population in crisis from the get-go (island populations are always fragile), the mature elm and chestnut trees succumbed to disease at the peak of their health and prosperity. Widespread large populations such as these are hard to kill. In fact the species in question have individuals who are still reproducing despite their brief lives. This leads to an interesting generalization about populations and infections: Infections (or infestations, or blights) rarely cause extinction. Large portions of populations may succumb to outbreaks, but there always seems to be some degree of resistance in every population, and some portion will survive.

There are American beech on our property in upstate New York. The beech, like the elm and chestnut, is threatened by a beetle that introduces a fungus into the tree. Beech trees were once abundant and huge, but by the middle of the twentieth century most adults in the populations of New York and New England were dead from disease. Medium-size trees survive, but usually in very remote settings, far from roads and surrounded by evergreens such as hemlock, which seem to insulate the beech from infestation. Again the population is not extinct—it has trees that reach reproductive maturity and produce offspring—but it is devoid of really large, timber-grade individuals. By the age of thirty years or so, beech trees succumb to the disease, whereas their ancestors lived to be more than two hundred years old. The last selective harvesting of timber on my property was around forty years ago, so these trees—obviously avoided by the loggers due to their youthful size—must be at least that old, but not much more. They are likely due to succumb within the next few years. I keep watch, but there's not much I will be able to do when the disease begins to show. I've seen slightly larger trees than those on our property, in forests nearby, that are in various stages of the disease. They tend to be broken, cracking, or bending to one side, with huge gashes in the bark, and showing few to no leaves in the canopy.

These species still exist, but not in a way that would be recognizable to eighteenth-century Indians or settlers. The small stands of trees that have some resistance cling on, though their life cycle has been cut short. It's important to reflect on resistance and infection in relation to our own population.

Our ancestors would be astonished to know that we no longer worry too much about infections or small cuts or abrasions. There was no such thing as a "minor" injury in the colonial era. Even if you had a doctor on hand to treat a wound, there was no effective treatment to stop any resulting infection. But our antibiotics cannot permanently protect us. Canadian researchers recently discovered a bacterium on raw squid, which was being imported into their country from Korea, that was resistant to carbapenems—the antibiotics of "last resort." The bacterium itself not particularly dangerous—the bigger threat was the possibility of it being ingested by a human and then passing on its DNA to that individual's gut bacteria.[185] With the increasing prevalence of infections that are untreatable with conventional antibiotics, this hypothetical human, no longer treatable with carbapenems, would die. It astonishes me that stories like this one, of populations and evolution, aren't leading the news every night. But instead news tends to focus on heroism and sensationalism.

As I write this, the Ebola epidemic in Liberia and Sierra Leone is in fact a top news story. The coverage, however, is skewed toward human-interest aspects—suffering of families, fears of spread in the United States, or most commonly, the bravery of the doctors. As soon as the human interest stories fade, I fear the news coverage will wane. The real story should be the expanding population of the virus (called by its generic name, *Ebolavirus*) and its doubling rate as seen in the infected people. The doubling rate is a rough measure of how fast a population is growing. It simply reveals the time it takes for a population to double its number of individuals. Since all populations grow exponentially at first (before their growth is slowed down by limiting factors), they follow a predictable pattern that can be depicted by a curve on a two-dimensional graph. One axis of this graph is a measure of elapsed time—years or generations—since the population began. The other axis measures number of individuals. All unchecked populations—those that are free to reproduce without limits, such as pathogens during an outbreak—show an initial, relatively flat trajectory for a period of time. But at some point the curve steepens suddenly, indicating very rapid population growth over little elapsed time. At this point populations are in the steep phase of their growth and are doubling ever faster. For diseases this often means that the pathogen population is out of control.

Ebola is being controlled by health-care workers, so the virus is not exactly free to expand. But the viral population experienced a nearly unchecked exponential growth for the first months of its outbreak. After being contained for nearly thirty-five years in Central Africa, the latest epidemic was first reported in December 2013, in countries with poor health systems, Guinea, Sierra Leone, and Liberia. In July 2014, as indicated by the number of infections, the viral population had begun to reach an initial steep point of its growth curve.[186] Deaths began to increase. By August one thousand people had died of the disease. By October that number had risen to three thousand in the three countries. By December it had surpassed five thousand. The doubling rate of the viral population continues to increase. This makes it progressively more difficult for health-care workers to manage the epidemic.

Ebola is most contagious (at its highest level of "virulence") when the host becomes a corpse. Virions are most abundant in the pools of body fluids that collect in the dead individual (blood plasma, lymph, saliva, urine, and so on). In order to manage this disease, careful stewardship of the pathogen's population is necessary. This comes from the workers doing the burying and containment of the infected corpses. As the doubling rate of infection increases, the workers may become overwhelmed. As of December 2014, the UN hoped to contain the virus by safely burying at least 70 percent of the dead in Sierra Leone and Liberia. The actual numbers are closer to 23 percent in Liberia and only 40 percent in Sierra Leone. This means that most of the diseased corpses are not being isolated, which leads to more infection of healthy unprotected bystanders. Many of these infections occur at burial rituals, based on long-standing cultural practices, in which family and friends touch the corpse before it is buried. If the knowledge of viral population growth and its effect on the rising death toll is not shared, decisions about rituals and isolation of bodies will continue to hobble the containment of this disease.[187]

Many of us have older friends or family members who contracted polio, whooping cough, or measles in the early to mid-twentieth century. These diseases of the past, and the Ebola outbreak going on currently, should be reminders that only through population management—containment of bacterial, viral, and other pest populations—can we hope to maintain the health of our own species. It is foolish to think we can vanquish any of them. Populations persist, and these dangerous ones are waiting in the wings for another opportunity to expand their range.

All biological and environmental threats to the well-being of our population require constant stewardship. Stewardship is not a simple one-step operation, such as pouring weed-and-feed herbicide and fertilizer on your lawn twice a year. In order to be good stewards of the biosphere, we have to foster an ongoing commitment to learning about other species. We need to find an answer to the fundamental question: What is the modern human ecosystem? before we can address how to manage the ecological interrelationships we have with other species. Stewardship refers to active management of those species and careful monitoring of ourselves in the context of a functional biosphere. I have faith that humans will—like most populations—persist through the hardships of the future. However, the grim possibility that our descendants will, like the beech and the chestnut, survive only in small, isolated populations, fearful of one another, merely eking out a meager living on their own small patch of land, goes against every hopeful bone in my body.

Homo sapiens is remarkably persistent. We've survived diseases that decimated continents, wars that wiped out entire generations, and natural disasters that obliterated whole countries. The worst human-caused impacts seem to kill off at most 50 percent of the population, as we will see below. Extinction, however, would be total. It is primarily associated with large-scale environmental climatological events. In the past, species have been wiped out by climate changes caused by unforeseeable events—meteor strikes, or vast amounts of atmospheric poison caused by volcanic activity. We are the first species that has been able to monitor and modulate the waste products that affect our environment. This means that we are also the first species that has a chance to reverse course and prevent hundreds of millions of climate-change-related deaths. In order to do this we need to learn to care about people, populations, and environments that seemingly have nothing to do with our own individual lives. A shift in focus, from individuals to populations, is the crucial factor that will bring this about.

I don't think that indulging in an overly emotional reaction to human tragedy is helpful. This does not mean, however, steeling oneself against compassion, or forgoing sympathy. There is plenty of suffering to be mindful of. Instead what is needed is a rational sense of empathy. Empathy means that we can understand another person's pain by extrapolating from a painful event in our own life. I am aware that I can't alleviate the pain for most suffering individuals in the world, but I feel motivated to do something about it. For instance, empathy motivates me to take action by raising awareness about the human condition in my songs, books, and lectures.

Most human suffering has identifiable causes. Our civilization learned how to avoid many of these dangers as it grew, passing on that knowledge to later generations, and we have now ameliorated many of the most insidious problems. Many diseases, weather disasters, and conventional wars, have, for the most part, been kept at bay in the Western world by modern technology, sanitation, and diplomacy. None of these advances is guaranteed to stay in place permanently. Populations are in a constant state of flux; and we are foolish to declare victory prematurely. This is especially true where we began this book, in the realm of conventional warfare, defining and trying to eradicate so-called enemy populations.

At certain times, such as during natural disasters, famines, or disease outbreaks, our human capacity for empathy is heightened. But sometimes caring should be balanced by a distanced analysis. It's this attitude I take when I consider that most newsworthy tragedies don't actually have a devastating effect on the standing size of the human population. We, as rational thinkers, have to acknowledge that our globe is interconnected as never before. Therefore our human population is nearly panmictic, which means that anyone has the potential to mate with anyone else on the planet. True, there is less likelihood of a villager in Africa mating with a farmer in Nebraska than of a Wall Street trader mating with a Chinese business executive. But there are plenty of examples of poverty-stricken foreigners from the Third World marrying Westerners and having kids who are brought up far from their parents' original habitats. We can thank intercontinental airplane travel for that.

Because of this, and because we are at such a steep phase of the exponential growth curve, the human population is at once relatively impervious to local "tragedies" and at the same time capable of producing more variations of new people to replace those who die in such catastrophes. In short, the rate of replacement is what a population scientist focuses on, and it's this perspective that I consider most prudent. However, most human-interest drama causes us to focus less on the rate of replacement and more on the loss of individual lives.

As one example, consider this. The largest single catastrophe from a weather-related incident was the Bhola cyclone of 1970 in Bangladesh (then called East Pakistan). This storm killed an estimated five hundred thousand people, many of whom were swept away by huge storm-surge ocean waves (aka tidal waves or tsunamis) and were never recovered. A similar tragedy happened more recently. In 2008 Burma (now Myanmar) experienced a cyclone called Nargis, that killed around 150,000 people. This cyclone sent a huge storm-surge wave twenty-five miles up the Irawaddy River and decimated towns along the way. Most victims were washed away in an instant. On the day after Christmas 2004, an earthquake brought equal destruction to residents in Indonesia. The quake, although registering as the third-largest Richter-scale reading in history

(9.3), was not the killer. Once again, it was the storm surge, this time caused by the shifting ocean floor that cracked apart from tectonic forces. A wave nearly one hundred feet high spread outward from the epicenter in the southeastern Indian Ocean. Within hours this wave struck land and nearly 250,000 people lost their lives in Indonesia, Sri Lanka, India, and Thailand.

These events are heartbreaking. When tens of thousands of people die in one short burst of "nature's wrath," we naturally feel that our population's future is vulnerable. But if we take the population scientist's approach to these events, we see that they don't actually predict anything about the health of our species at large. In fact that might be the disconnect, and the greatest challenge to our ability as stewards of the planet. How can we manage our species if we get caught up in these horrific disasters? I'm not sure the answer is easy, but certainly the facts are sobering.

In Japan the birth rate in 2011, the year of the 3/11 Tohoku earthquake and tsunami,[188] was about 7.3 per 1,000 people. The disaster may have removed twenty thousand people from the Japanese population, but from new births, roughly twenty-six thousand were added to the population every month that year. So in one sense, though a very callous one, it took only about twenty-two days for the Japanese population to restore its population size . Given the rate of population growth in the northern Bay of Bengal at the time of the 1970 Bhola cyclone, it took the residents of that region only two months to replace the numbers of those killed in the storm with new births.

So, if natural disasters don't do much to dent the human population, what about warfare? Certainly we've heard plenty of stories about the vanquishing of previous civilizations by dominating armies and occupations. And in fact the numbers of casualties from modern warfare are staggering. Until the twentieth century those killed in individual battles by "conquerors" in history made up relatively small numbers, on the order of tens of thousands at worst—similar to some of the natural disasters just described. A notable exception is the Mongol conquests of Genghis Khan in the thirteenth century. Inspired by territorial imperatives, Khan and his men slaughtered nearly forty million people, who stood in the way or resisted the spread of Mongol culture. Considering that the world's population was only around 400 million at the time, this means that Genghis Khan was responsible for killing one in every ten people alive! This may well have caused a population bottleneck, eliminating a sizable portion of the standing diversity at the time. Khan and his men, however, also raped many of the women along the way, thereby unwittingly assimilating the very populations he hoped to vanquish. And through it all, the world population continued to grow.

By the middle of the nineteenth century, there were roughly 1 billion people on the planet. During that century in China, the Taiping Rebellion saw roughly twenty million die from warfare, while more than five hundred thousand died in Spain in a series of civil wars. Some 750,000 people perished in the American Civil War. These wars merely set the stage for what was to come in the twentieth century.

The most brutal wars came as the twentieth century matured. Roughly fifteen million people lost their lives as a result of the First World War, and since the issues that caused it were never successfully laid to rest through diplomacy, World War II rose up amid its ashes and caused the greatest loss of life ever witnessed from human conflict. Between sixty and eighty million humans died during that war in the years 1939–45. In China and Russia, civil wars took place between the two world wars, resulting in the deaths of roughly twelve million people. It's normal to wonder what effect this loss of life had on our population structure[189] (I've always maintained that the world lost its mind in the two or three decades before I was born.)

At the end of World War I Lenin called for a European revolution based on socialism and meant to establish communist communities of peasants and workers. At the same time Woodrow Wilson proclaimed that a League of Nations, headed by the United States, should be based on democracy, not communism (the United States eventually failed to join). Russia and the United States, allied in the fight over Europe, harbored opposing, incompatible ideologies that remained intact through another global conflict, World War II. At each step of the way violence made citizens miserable or worse, while the promises that came from the war's architects always fell short of fulfillment. "Victory" was felt only in the relief of misery, but not in the vanquishing of some enemy population.

Nearly half of the sixty to eighty million people who lost their lives in the Second World War were civilian. The end of the war brought "peace" in name only, as Russia and the United States had vast ideological differences on how the world's states should comingle. It took almost fifty years, but in the end the Soviet system gave out (in 1989–90), and today democracy is not only the mainstay of the Western world, but it has displaced imperial and dictatorial regimes, and has been an intrusive partner of socialist governments worldwide.

But socialism wasn't vanquished, either as an ideology or a policy. Congress is currently debating the very same issues that Russia and the United States debated in 1918 (namely the role of centralized government in provision for and procurement from its citizens). In other words the global wars, the carnage and suffering, of the last hundred years ended without an ideological conclusion to bring disparate populations together. It's as if warfare in the twentieth century and even today were just part of a never-ending philosophical debate. Yet despite all these senseless deaths from war, the global population of humans continued to increase, from about 1.6 billion in 1900 to about 2.5 billion in 1950, to today's 7.1 billion. It would take a lot more than hundreds of millions of deaths from warfare to vanquish our population.

Unlike typhoons, earthquakes, or other natural disasters, war is predicated on failed diplomacy. It can be prevented! But, more important, war can be understood. That is to say, the way we treat human warfare in our historical narratives can be rewritten to be more consistent with how we explain the evolution of organisms. Populations, by their nature, come to coexist with much difficulty; many, many individuals will die in the process of assimilation. War doesn't have to be this way because we, unlike animals and other species, can put ideology aside if we value the lives of individuals and the potential benefits they bring to our population's future. No other species has the ability to do this. We should be able to foster a more harmonious cultural mélange using policy and reason, without the extreme violence and upheaval caused by conventional warfare.

The brutality of human wars is only one of two major hurdles we need to clear in order to achieve a satisfying shift in focus away from individual suffering toward that of true population stewardship. The other is the burgeoning environmental crisis we are busily creating for ourselves. If we can learn to manage everything from wild species to microbe populations, we should be able to create a global, sustainable environment for ourselves.

All extant organisms have the tendency to coexist. Extinction takes place despite, not because of, this coexistence. So it's crucial that we recognize the following: Nearly all mass extinction events correlate with some kind of atmospheric or oceanic disturbance. Our own species is not immune to extinction; if we wish to persist we need to look at our predicament from a scientific perspective, just as we might consider the plight of any endangered animal. Or, as a biologist would say, we need to figure out a way to preserve our longevity as an evolutionary taxon.[190] The recognition that our atmosphere and oceans are being fouled by our own industry means that we are not only unscientific but supremely stupid if we don't do everything in our power to clean them up. This will, as a first requirement, entail widespread education about the atmosphere and the oceans, bolstered by strong pollution laws and the spread of environmental ethics.

Earth history has to play a central role in the new narrative of the twenty-first century. All discussions of sustainability ultimately come down to expectations with respect to the environment. The rocks beneath our feet record a long history of environmental change and contain correlated atmospheric signatures. They tell us that mass extinctions are accompanied by climatic perturbations. Human hunting or overuse of natural resources only exacerbates the "natural" process of extinction, especially if the species in question is already out of equilibrium with its environment and its population numbers are dwindling. If we are serious about being stewards, therefore, we have to understand our role in disturbing the environment. This ultimately boils down to what we consume, what we throw away, and the by-products created in the process.

In early Earth history it was an easier equation than it is today. Those populations that first inhabited the planet were cyanobacteria or something closely akin to them, around 3.5 billion years ago. They consumed water and CO_2 and produced oxygen as a by-product. At first, this oxygen was sequestered in sediments that rusted (oxidized) and formed great deposits called banded iron formations (BIFs). Also called "oxygen sinks," these sediments were soon unable to absorb the excess oxygen produced by rapidly proliferating populations of photosynthetic organisms, and it began to pervade every part of the primitive oceans. So much oxygen was produced, in fact, that it created a crisis in the atmosphere and oceans: too much oxygen. Oxygen is toxic to most organisms, and much of evolutionary "innovation" has been devoted to dealing with this reality by chemical means called "antioxidant molecules."

Had not the earliest life-forms evolved a means of protection, they would have succumbed to oxidation poisoning from their own waste products after sedimentary oxygen sinks had been used up. Today the cells of many species are surrounded by walls made of cellulose (plants) or peptidoglycan (bacteria) as a protective layer against the damages of atomic dismantling from oxygen. More effective yet, antioxidant molecules were an early evolutionary innovation of photosynthetic cells. They act as cellular oxygen sinks. In the case of animal cells, these substances are contained in small chambers inside the cytoplasm called peroxisomes. Peroxisomes contain hydrogen peroxide, a molecule found in almost every cell that comes into contact with oxygen. The ubiquity of this molecule suggests its crucial status. It likely took hundreds of millions of years to evolve, but hydrogen peroxide contained in peroxisomes was the result of the atmospheric buildup of oxygen. The evolution of this oxygen-absorbing organelle may have begun around 2.1 billion years ago, and it might have been the defining innovation that led to the eventual evolution of all the familiar kingdoms of life on the planet, including fungi, plants, and animals.[191]

Peroxisomes may have been the single most important player in the early Earth's oxygen toxicity crisis, which began about 2.1 billion years ago. Microbes that contained these organelles (certain primitive eukaryotes) or those that could protect their cell contents by surrounding themselves with some kind of cell wall (peptidoglycan in some prokaryotes) were safe. But most cell biologists believe that there was a mass extinction at this early stage of the biosphere. The majority of cells and primitive colonies could not withstand the atmosphere and aquatic environments because they were quickly being poisoned by oxygen from photosynthesis. In other words these early prokaryotes were the first populations to establish the tradition of polluting the atmosphere due to careless overabundance.

Although the buildup may have been gradual, eventually all organisms on the planet became extinct except for those that had some kind of antioxidant mechanism to contend with the toxic environment. Some scientists posit that the first mass extinction on the planet was in response to this "great oxidation event"—the buildup of oxygen from photosynthetic microbes—during the Precambrian. There is a certain amount of irony in the fact that even the very first living cells on the planet eventually caused their own extinction by changing their environment.

We don't have to look that far back in history to see life-threatening atmospheric perturbation; we are experiencing such an upheaval today. But it requires a shift in focus to recognize it. Just as in the stupidity of seemingly endless cycles of warfare in human history, we, like all organisms of the biosphere, blindly fall victim to our own exuberance of environmental overexploitation.

Global warming has become a politically charged discussion that divides scientifically minded citizens from conspiracy theorists and their advocates. The former, led by an international group, the IPCC,[192] maintains a staunch position that CO_2 from human activities is drastically affecting global warming. Hundreds of top scientists, performing ongoing research in chemistry, geology, atmospheric sciences, and biology, issue frequent reports on the state of our atmosphere. This group is counterbalanced by citizens who view science and global warming as a conspiracy to mislead the public. Many of those fueling the flames of doubt against scientists are industrialists, representatives from the coal and gas industries and the billionaires who make their fortunes from the unrestrained combustion of fossil fuels.[193]

There is no doubt that carbon dioxide in the atmosphere is correlated with Earth temperature and it has been so for most of Earth history.[194] The degree of warming caused by humans in the current rise in atmospheric CO_2 is debatable, but certainly more of it in the atmosphere means adding to, not detracting from, our woes of higher global temperatures. Whenever glaciation has predominated on our planet—and we are in a relatively cool phase currently, although much warmer than it was 12,800 years ago—there is an associated historical signature in rocks or ice cores that indicates low levels of CO_2 in the atmosphere (lower than around 500 parts per million). In 1860 the atmosphere contained roughly 280 ppm, and today it has risen rapidly to around 380 ppm (about a 36 percent rise). This is due to industrialization during that time span.

Whether atmospheric CO_2 causes global temperature to rise directly—as in its reflection of infrared radiation—or indirectly as a correlate of other factors, there is no denying that it is a potent greenhouse gas that needs to be closely monitored and regulated by environmental policies. If you're not alarmed by the catastrophic weather-related disruptions to human health and welfare that are reported daily, then perhaps you should be alarmed by the drastic tornado activity in the Midwest (where warm, wet air of the midcontinental United States collides with cool air coming in from the West), or the nearshore coastal communities that are progressively becoming inundated at the current rate of 1.5 inches per decade, or the melting of glaciers in Antarctica and Greenland, or the measurement of global temperature in August 2014 as the highest ever recorded.[195] These issues reveal that we are in the midst of a global climate change. If our ethical objective is to be good stewards of the planet in an effort to avoid human extinction, then our greenhouse-gas emissions should take center stage in all economic and social discourse. In other words our human population wars are ultimately at the mercy of the climate. If we successfully shift our focus to the most pressing environmental concerns—oceanic and atmospheric health—we will alleviate strife between all groups because we will have to cooperate to overcome such global-scale problems.

As I've shown throughout this book, Earth's history is an episodic sequence of population wars. We are, as Shakespeare famously wrote, actors for whom "all the world is a stage." It's just that none of us gets that much time in the spotlight in the drama of evolution. Our individual lives are so short as to be almost insignificant—even the lifespan of a species is relatively brief in relation to the time frame of geologic history. Every epoch of Earth history is defined by species of fossil organisms that failed to live on into the subsequent period. For instance, the Frasnian-Famennian boundary divides the Late Devonian Period in Earth history (about 372 mya). On either side of this stratigraphic boundary are fossils representing vastly different ecological communities that were relatively intact for about ten million years. During the Famennian Age, the younger of the two, tetrapods first evolved, a distant, four-legged human ancestor we share with amphibians. No true tetrapods existed prior to this in the Frasnian. Of the fish that occurred in the Famennian, all had ancestors in the older Frasnian Age, but those ancestors were all extinct. They didn't make it to live among their descendants in the Famennian. There was nearly total loss of jawless fishes, the most primitive members of our subphylum, Vertebrata.

I point out this boundary not because of its significance, but rather to use it as a representative of hundreds of similar such formal boundaries recognized by paleontologists throughout the fossil record. Every major division in the long record of Earth history is subdivided by these smaller boundaries. The thing to keep in mind is that they are all defined by faunal turnover (or floral turnover, when you consider plant communities). That means that every ten million years or so there is a significant-enough difference in the communities of species alive at the time to warrant a new stage in the formal classification of evolutionary history.

Because of this fossil evidence I prefer to see evolution as a sequence of generally disconnected episodes of coexistence rather than as a smoothly branching, graceful "tree of life." The episodes are distinct; the species that coexisted during the deposition of one geological stratum are different from those that coexisted in the subsequent geological era. Yet there is a connection, even if the characters have all changed from one episode to the next. In all of Earth's historical stratigraphic subdivisions there were communities of species that depended on one another and reached some sort of equilibrium, a state of ecological balance. It can be disrupted, such as in times of environmental crisis, but eventually new equilibria are established and life goes on for those populations that are lucky enough to make it to the next stage.

Each tick of the geological time scale brings countless extinctions; in fact extinction in Earth history is as common as warfare in human history. It's these repetitive sequences of tragedies that we use to calibrate life on Earth. Consider one of countless examples, from a slice of Earth's rock record that I studied as a graduate student, the Ordovician Period (485–444 mya).

Climate change affected the Ordovician biosphere in a number of ways. Glacial sequestering of water (freezing on continents) meant there was less water in the ocean basins, which caused exposure of the continental margins. These were, as today, nearshore reefs, precisely the areas of the greatest biological diversity in the ocean. At this time in Earth history there was no terrestrial ecosystem as we know it today. Neither plants nor animals had evolved on land yet. Habitats of nearshore marine organisms, once bathed in shallow water that received nutrient runoff from the continents, became flooded with harsh sunlight, exposed to cooler temperatures, and simply dried up. By the end of the Ordovician, there were 85 percent fewer species than were alive in the middle of the period. Some of the most popular fossils for collectors, trilobites (those flattened, insectlike creatures with greater than twenty pairs of jointed legs), lost many of their species, as did the clamlike group called brachiopods, and the clams themselves, the Bivalvia. Tinier creatures than these, viewable as fossils only with microscopes, which resemble minuscule branching woody plants—but are actually animals—the graptolites and bryozoans, also died out en masse. Perhaps the most familiar organism to go extinct at the end of the Ordovician were certain types of corals.

Coral reefs are the most biologically diverse ecosystems in the oceans. They are nearshore communities of species that live on, inside, and around secreted skeletons of stationary animals. Usually when you look at a coral in a museum or collection, you have a sample of only its skeleton, since the soft tissue has decomposed. The first thing you notice is that it's composed of a hard, limestone-like material that is pockmarked with cavities. Inside each of these cavities is where an individual animal lived. Corals are colonial populations all living together in this combined skeletal structure. Over time new generations build up on top of previous ones. Billions of individuals combine to form the framework of coral reefs, and a single fused-together colony can extend over hundreds of square miles.

Mass extinctions in Earth history have always destroyed the world's reefs. In the Ordovician Period, the dominant reef species (aka, the framework species) were stromatoporoid sponges, flat matlike organisms with multiple layers of sequentially deposited skeletal calcium carbonate (calcite or $CaCO_3$). Another reef framework builder at this time was an alga called *Receptaculites sp.* This organism looked like the head of a sunflower, and like the sponges, also deposited calcitic skeletons. These were accompanied by a group called Rugosa, or "horn" corals. Much different in anatomy than our corals today, they performed the same ecological function—catching tiny prey using specialized stinging cells or trapping free-floating plankton in the nearshore environment, excreting nitrogenous waste, and building calcitic skeletons that contributed to the framework of the reef.[196] All these groups formed the framework of a huge and complex biological community. When they went extinct due to climate change, glaciation, and the disappearance of their nearshore environment at the end of the Ordovician, all the other species that depended on them also went extinct. When reefs die, it is not simply the loss of a single coral species, it is the erasure of an entire web of coexisting animals, plantlike algae, protists, and bacteria.

Fossils don't provide the only window into mass extinction. Thanks to geochemistry we can measure the signature of ecologically significant elements in rocks that indicate changes in the primitive atmosphere. Specifically, carbon can be traced in Earth history as never before. We can now put a correlative atmospheric chemical stamp on the fossil record. Elemental variations (called isotopes) tell a story about the environment that accompanied the mass extinctions.

In each mass extinction, there was a rapid change in atmospheric carbon.[197] Most commonly this was due to rapid fluctuations (usually an increase) in atmospheric CO_2 levels. In fact, after the extinction of reefs during the "Big Five" mass extinctions, the Ordovician being one of them, it took roughly ten million or more years for new reef communities to become established. These are known as "reef gaps" in the fossil record, and they indicate voids of biodiversity inflicted by the environmental aftereffects of climate disruptions. Carbonic acid, created by high levels of CO_2 in the atmosphere, enters the reef environment as rain. If sufficiently high levels are encountered, the ocean can't absorb or "buffer" the acid, and the water chemistry is affected (lowering the pH level). This is just one of several environmental factors, correlated with the carbon cycle, that cause reefs to die. Algae that secrete calcifying cement are also affected by acidification. Without the physical process of mineralization created by organisms all working in synergy, the reef community is doomed. Reefs, like forests of the ocean, are, and always have been, the bellwethers of ecological health, and today's reefs are showing signals of stress.

Recent monitoring studies have found that reefs worldwide show between 27 percent and 35 percent dead coral polyps (the tiny animals that live on and secrete the framework of the reef).[198] Meanwhile, nitrates from agricultural runoff end up in the nearshore and cause a huge increase in algae. These populations encrust on top of the reefs and destroy the living polyps. Furthermore, warmer surface waters (from increasing greenhouse conditions), cause "bleaching" of the corals, which eventually destroys the reef's vitality and causes depletion in the entire community of reef species. This is particularly relevant to the theme of this book because it shows how environmental perturbation causes a breakdown of symbiosis.

Arguably the most important element in today's reef ecosystems is a genus of photosynthetic algae (or protist, depending on the classification) called *Symbiodinium*. This genus is a member of a larger group of single-celled, mostly free-swimming algae called Dinoflagellates. Individuals of the tiny single-celled *Symbiodinium* live in huge numbers, up to one million of them in a single cubic centimeter of coral polyp tissue. They form a mutualistic, symbiotic relationship with the corals, and contribute to the growth and secretion of the reef's carbonate framework.

Beginning their lives as free-swimming organisms, these tiny dinoflagellates are taken up through phagocytosis by individual cells of the coral tissue. Once inside the coral tissue, *Symbiodinium* begins its mutualistic relationship. Both inorganic and organic molecules are exchanged between both partners, host coral and mutualist dinoflagellate. This means that both populations continue to grow, replicate, and thrive in the presence of the other. Each partner in the symbiotic relationship brings valuable resources to the other. The lovely colors of corals are due to varying amounts of photosynthetic pigments contained in their symbiotic dinoflagellate population. Corals are not photosynthetic organisms, but they derive benefits from the sugars and other compounds produced by photosynthesis due to their symbiotic dinoflagellates.

When waters become polluted, or when temperatures rise in the surface layers of the ocean, corals become stressed. The first action taken under such conditions is an expulsion of their symbiotic partners. It's almost as if Poseidon puts out a clarion call for all organisms to "save yourselves" in times of ecological distress! In reality, bleaching is due to the lack of symbiotic algae, and this indicates that the coral is damaged. What's worse, expelling the dinoflagellates only makes them less healthy. Corals begin to starve when they expel their dinoflagellate residents. Bleached coral is not dead per se, but it is not actively growing or thriving either. Numerous reef species that depend on coral for their sustenance, such as "grazing" fish and mollusks that eat the polyps, disappear or begin to starve. In short the disappearance of the dinoflagellates sets off a chain reaction that depletes the entire community of reef species. The best estimates indicate that today's reef communities are disappearing. Some of the reefs in the Florida Keys have lost 90 percent of their coral population in the last forty years.[199]

Less than a call to action for reefs specifically, this data is most alarming because of what it says about the environment that we have created. Using the history of extinct reefs as our guide, we can already see the beginnings of a modern-day mass extinction upon us. Shifting our focus toward that of population stewardship might save the reefs, but more important, it might be just the thing to save our own species from this mass extinction.

The time frame of extinction is not easy to grasp. Only very rarely do examples like the dodo present themselves, which, as mentioned, went extinct in a relatively short period of time and was recorded in journals as it happened. Yet extinction is a fact of the fossil record. This record is imperfect, which means that it's too coarse grained to reveal the slowness of population decline. Think of the fossil record this way: If an inch of strata takes ten thousand years to deposit (as is reasonable to assume in layers of mudstone), that means that finding a fossil, say a snail shell, in one of those layers reveals one species alive during that particular span of time. If no more fossils of those snail shells are found in the overlying layers, all we can say is that somewhere in the span of ten thousand years, the species became extinct. Maybe it was a slow death of the population, maybe it was instantaneous, but because the sample size of the fossils is so small, and the time span between strata so long, we cannot pinpoint the exact time or rate of the extinction. Often, due to the vagaries of the fossil record, we deal with millions of years rather than tens of thousands.

So, if we consider the coarseness of the fossil record, we have to accept that extinctions might be slow. We know, for instance, that our own species has existed over a time span of roughly two hundred thousand years during the Pleistocene Epoch of Earth history. Since that time there have been numerous notable extinctions of large mammals that comprised a community often referred to as the Pleistocene megafauna. The La Brea tar pits in Los Angeles, one of the richest fossil sites in the world, contains a great sampling of many of their remains. One of the most important aspects of the La Brea tar pits is the evidence that humans lived alongside these giants of the past. Among the strange species is the giant ground sloth, genus *Paramylodon*. This beast weighed fifteen hundred pounds, could reach up to branches nearly twenty feet high, and ate rough vegetation and leaves. Nothing like it exists today. Our ancestors also lived alongside the saber-toothed cat, genus *Smilodon*. Larger than any lion alive today, this ferocious predator hunted huge elephants (mastodons and mammoths) that roamed North America and Asia. Both the saber-toothed cat and the elephants are now extinct. The short-faced bear, genus *Arctodus*, is another veritable La Brea giant.

Standing five feet at the shoulders (when on all fours), this bear weighed nearly one ton and was as ferocious in its predatory behavior as any grizzly alive today. But, like its cohorts in the Pleistocene megafauna, it is not to be found anywhere today. Likewise for other huge mammals with whom our ancestors shared the late Pleistocene landscape: the ancient camel, genus *Camelops*, the huge dire wolf, genus *Canis*, and the seven-foot-tall ancient bison, genus *Bison*. This is just a mere sampling of the many species that made up the Pleistocene megafauna, but all are extinct today. In fact all large mammals of the Pleistocene megafauna died out by eleven thousand years ago. Our own species, however, has continued to expand its range and increase its population size since that time.

A paleontologist of the future might discover sediments that contain these huge fossil mammals commingled with bones from modern humans. The sediments filling the Los Angeles Basin are slowly piling up with each rainstorm that washes down the canyons. All the debris from modern humans, including their remains from the nearby Hollywood Forever Cemetery, might be washed together with the fossils from the La Brea tar pits into one great sedimentary unit somewhere just offshore, perhaps in Santa Monica Bay. This sedimentary layer, containing a mixture of our modern human remains with bones from La Brea tar pits, might get quickly covered by sediments washing down off the Santa Monica Mountains, whose canyons empty directly into the sea.

In this case of a hypothetical mixture of "young fossils" from the La Brea tar pits with modern bones of today, there would be no recognizing the megafauna as distinct from modern man. The stratigraphic horizon would be a jumble of bones that would all seem to have lived contemporaneously. The paleontologists of the future would have no other data than those allowing them to conclude that we modern humans lived alongside the Pleistocene megafauna. There is truth to this conclusion because we already know, from other fossil evidence, that humans lived alongside the megafauna. But what this possibility says to me is that this stage of Earth history is still ongoing. The Pleistocene extinction might still be under way, and we are living to witness it.

The endangered large mammals of today, the polar bear, cheetah, elephant, rhino, or timber wolf, for instance, when viewed in the light of the already dead megafauna, and coupled with the data from the declining vitality of bleached corals, tell me that there is a good reason to view extinction as a long-term phenomenon, and consider that we might become a very widespread and abundant fossil as part of this "sixth" mass extinction.[200] The fact that we continue to increase our population, however, is both a blessing and a curse. It means that we are doing something right in the midst of all this extinction of large mammals. But it also points out how crucial it is to heed the examples I've used throughout this book to remember that humans are subject to the same laws of population growth, equilibrium, and extinction as other species, if we don't actively manage our evolution.

The mention of managing evolution in humans leads immediately to the fear of eugenics, the misguided attempts of geneticists, particularly in the early twentieth century, to change the human race through breeding. Led by Charles Darwin's cousin Francis Galton, an otherwise gifted mathematician and statistician, an entire cadre of intellectuals, geneticists, and politicians agreed that one of the great imperatives of mankind should be the elimination of harmful and less-desirable genes from our species. "I . . . maintain that it is a duty we owe to humanity to investigate the range of that power [of breeding in humans], and to exercise it in a way that, without being unwise towards ourselves, shall be most advantageous to future inhabitants of the Earth."[201] There is an undeniably hopeful tone to Galton's mission. Just as the practice of animal husbandry changed livestock into improved races, he and his followers believed that through scientifically rigorous social programs, the human race could be improved. Since Galton's main focus, originally, was related to breeding for genius, it should come as no surprise that his followers devised plans, not altogether pernicious, to distinguish between those of low and high mental ability.

Sadly, this mission, begun in 1883 with Galton's coining of the word "eugenics" (meaning "of good stock"), devolved into a misguided scientific agenda after genes were discovered around 1900. The pervasive belief among geneticists of the early 1900s (who at the time knew nothing of the genetic complexity of human intelligence) was that "feeble-minded" people carried bad genes. If such people could be prevented from passing those genes on to future generations, they thought, the stock of the human race would improve. So, much energy and expense was wasted in measurement of races, classes, occupations, and abilities, in an attempt to quantify their degree of intelligence. In the United States this led to a tragic program of forced sterilization of the "feeble minded," justified by law, first in Indiana in 1907, and eventually ratified by the Supreme

Court in 1927 (*Buck vs. Bell*).[202] Called "genetically unfit" under these laws, an estimated 65,000 people were sterilized without their consent in the United States in the first half of the twentieth century.

The idea that intelligence could be reduced to a single, measurable, quantity, based in the genes, was taken to extremes in those days, and scientists—particularly social scientists—applied the idea to an outdated scale of human races from the eighteenth century, placing Caucasians at the top.[203] In more recent times the reflection of this racism, based in eugenics, has been espoused by those who promote the IQ test as the best way of measuring human intelligence. Intelligence is dominantly affected by culture and its impact on development of the brain, that is, neuronal selection. Unlike the genes, intelligence can, therefore, be changed dramatically during a person's lifetime through education and new experiences. Thus intelligence is not a measurable genetic variant that is useful in characterizing racial differences in humans. As the geneticist Richard Lewontin succinctly noted: "The genes for intelligence have never been found." [204]

The prevention of human breeding through sterilization is now considered a form of genocide, and it is condemned by all member nations of the UN. The belief in "improvement" of the human race is not generally considered possible, or desirable from an evolutionary perspective. First, it's difficult to think of a trait that is universally considered "bad." Second, most traits are linked to other traits, so getting rid of a so-called bad trait in a population might also affect the proliferation of a beneficial linked trait. But third, and most important, whatever we view as a bad trait is arbitrary, and should instead be seen as a variant. Variation is the raw material of natural selection. Who is to say that a bad trait today will retain its "badness" in the future, when conditions change? Even if we could breed "bad genes" out of our genomes, in a very real sense we would be short-changing future populations by removing variation, and thereby lowering their potential for evolutionary change.

With the outrageous failures of the early geneticists, one would think that the science of human improvement might have died a long time ago. Not so. Instead of government policies on breeding, however, today's eugenics has taken on a new name: personalized genomics. Today individuals are given choices about reproduction that allow them to feel some degree of control over the future complexion of their family. These screenings have become commonplace for cancer and other diseases.

For less than one hundred dollars, any couple planning on having children can get a genetic screening, which is a complete readout of both partners' DNA sequence. Within this sequence, specialists recognize gene variants (aka alleles), some of which might be detrimental if combined in a certain way with other gene variants. Through decades of accumulating genetic data, hundreds of genes are now known to correlate with various cancers and other diseases. If a specialist determines that both partners of the couple are carrying a disease gene variant, but not showing any signs of that disease, it is possible that they could remain healthy but give birth to a child who develops the disease. In such a case the couple might decide to forgo having a child.

This doesn't sound like evolution management, but it is. If a couple decides not to pass on their genes, the future population will not inherit any of their genetic variations. Since our population is so huge, however, individual actions like this will probably not even make a dent in the future of our species unless they're widely adopted. We cannot revert to the legally implemented genetic policies of the early twentieth century, even if we did universally agree that cancer genes are bad. Government mandates on breeding in humans are just too unsettling ever to be viable again. But I believe we can still manage the evolution of our species by paying close attention to the environment. It's less of a question of population genetics and more a question of determining the limits of environmental parameters under which the human organism thrives, and then committing to maintaining ourselves within that range.

I'm a child of the seventies. As a kid I wore T-shirts that said, "Save the Whales," "Save the Pandas," or simply "Save the Planet." My focus was on saving other species because we—*Homo sapiens*—didn't seem at risk. I'm not so sure about that anymore. Now I think that when we talk about "saving the planet," what we really need to think about is "saving ourselves as well." I have friends who are such committed environmentalists that they think the best possible outcome is an Earth that has been wiped clean of human beings, where the natural world has returned to some kind of prehuman equilibrium. I don't understand this idea; I don't see *Homo sapiens* as being some kind of irredeemably flawed species. My friends who embrace these kinds of extreme ideas are against the destruction of anything in the natural world—other than humans. Their point of view, which I understand, is that enough of nature has already been destroyed. They believe that instead of indiscriminate destruction we must now embrace indiscriminate preservation. The problem with this philosophy, as pointed out earlier, is that it does not take into account the fact that we live in a world that is already disturbed.

Disturbed areas are vulnerable; and in order to maintain any kind of population balance you have to be willing to manage actively what gets to live and what has to die. Anyone who has tended a garden or mowed a lawn has done so without thinking twice about it. I do it routinely as a landscaper on my property in upstate New York. In fact some of my friends are surprised to see that I don't hesitate to use the controversial weed killer Roundup to control unwanted invasives. This to me is less about supporting some supposedly conspiratorial multinational company (Monsanto invented Roundup), and more about population management or stewardship. Applying Roundup responsibly, in limited areas, during alternate growing seasons, is an effective and safe way to kill weeds and prevent incursions from invasive species. Almost everyone, from organic farmers to florists to national park rangers to golfers, agrees that invasive plants bring with them invasive animals, and all of them are bad.

We all have inherited a disturbed world in which our forebears have artificially created and exaggerated areas of ecological imbalance. Oceans are in crisis because of pollution, global warming, overfishing, rising acidity, and the great gyres of trash in the Pacific. Substitute a forest for an ocean and it is the same story—there is no going back to an older version of some idealized, natural, prehuman world. The only way is onward. We have to be active stewards of our environment and operate in a logical and rational way. I am incredibly sad that my large beech trees die at such a premature age, but that is the new reality of the arboreal world, and I have to accept it. Instead of lamenting the fact that I'll never see a stand of healthy chestnut trees, I have to manage the populations that have persisted in their stead.

As a landowner, I have a say about what happens to my small patch of eastern hardwood forest. Legally I own that right, as well as the rights to any minerals underneath the forest. What I don't own, in any logical sense, is the species themselves, or the ecosystem in which they participate. I have taken it upon myself, through a sense of ethical stewardship, to preserve as much of that ecosystem and keep it untrammeled as possible. Our elected officials and every citizen who consumes natural resources need to take the same responsibility. If they can't do this by understanding the implications of science or by reflecting on facts of evolutionary history, then at least they can achieve this through understanding that they are dependent on other populations for their own well-being.

Having some control over the environment could rightly be seen as humankind's greatest achievement, greater than walking on the moon, greater than the Internet, greater than modern medicine. I say this because dealing with environmental hazards is the one thing we share with every other population that has ever existed in nearly 4 billion years of Earth history. The earliest life-forms created atmospheric pollution that nearly destroyed them. Through the evolution of antioxidant molecules and other "strategies," those hazards were overcome, but it took nearly a billion years. In a sense we are still contending with the challenges of the earliest population wars. From Earth's most primitive organisms to us is a long, circuitous, line of descent, and we have ended up in the same predicament. Like our bacterial predecessors, we have unwittingly released poisons into the environment from our activities. Also, as seen in the early cyanobacteria, the effects of this pollution have altered the other species with which we coexist. For the first time in the history of life, however, we can break free from the unwitting poisoning of the planet because we have an evolved trait that they lacked: consciousness. Through the monitoring and control of production, consumption, and disposal, we can manage the evolutionary constraints to which other species must adapt.

Charles Darwin recognized that adaptation entailed two components: survival and reproduction. In order for evolution to occur, populations have to show variation of its individuals in their ability to contend with the environment (survival) and to leave more offspring (reproduction). For the first time in the evolutionary history of life, our species now has the ability to manage each of these components to some degree. We can sequence the genome of any organism, including our own, use biomolecules to cut out sections of DNA, splice them together, and manage their gene products (beneficial or harmful proteins). We can invent machines that create few to no toxic by-products for the environment. We can control other populations, from pathogens to household pests, by killing them, displacing them, or immunizing ourselves, as never before. We can even modify the genetics of food populations upon which we subsist (from fish to plants and everything in between). But what we can't seem to do is shake the pervasive belief that a God or some other force is in charge of our ultimate destiny.

Those who believe that God's hand guides everything in the universe are in the same nihilistic boat as scientists who believe that there is nothing we can do to prevent our imminent extinction. They both leave the destiny of our species in the hands of something over which we have no control—either a deity or a grim statistical probability based on other mammals' longevity. I don't identify with either of these groups. I'm a naturalist, and therefore I find solace in the facts of an evolutionary worldview, which teaches that nothing is predestined by a deity or by a static system of genes interacting with the environment. The biosphere is in constant flux. I believe that we have some control over both of the major components of evolution, survival with respect to the environment, and reproduction of our own as well as other species. This is all the validation I need to come to the conclusion that humans can be effective stewards of the planet and thereby guide, however coarsely, our own microevolution.[205]

My view still leaves room for debates over ultimate fate. Throughout the course of Earth's history, as we have seen, there are unforeseen tragedies that affect all species (mass extinctions). Despite much biogeochemical research and alarming preliminary data on mass extinctions, it is at least possible that we have no control over these catastrophic biocrises. If that is the case, and we cannot predict such large-scale depletions of biological diversity, I see no reason why a religious person's position that "only God knows when the next mass extinction will occur" should not be considered at least equal to that of the scientist who claims "we cannot predict when the next catastrophic event will occur." If, however, the data continue to accumulate showing that past mass extinctions are correlated with widespread and rapid swings in CO_2, then we have another reason to abandon faith in all supernatural forces and accept our role as stewards of the environment. Knowing where we have control and admitting where we don't are a guide to further research. Relinquishing all responsibility to some supernatural entity—fate or God—is hopeless.

Back home on the farm I see incremental improvements in technology that give me reasons to be hopeful. My old tractor is a 1962 Allis-Chalmers D19, and it burns gasoline. At the time of its manufacture very little concern was paid to hazardous emissions. Even though it has no emissions technology built into it, I don't use it enough to make a significant dent in the atmosphere. I ride it only rarely. My newer tractor is a 2015 CaseIH 75C with a diesel engine that meets or exceeds the clean emissions standards set by the California Air Resources Board (CARB) and the U.S. Environmental Protection Agency (EPA). Our family car is the cleanest-burning vehicle on the road, with a high-efficiency diesel engine that give us forty-three miles to the gallon and whose exhaust is mixed with diesel exhaust fluid (DEF), which turns poisonous nitrous oxides (from normal combustion of fuels) into harmless, inert components of the atmosphere (water and nitrogen gas). As mentioned earlier, our house uses less water, is heated by less energy input, assisted by better insulation, and contributes less biological waste to the environment than any comparable house built within the last fifty years. All these are hopeful signs that incremental improvements in technology, coupled with greater public awareness, strict controls on industrial production, and enforcement of rigid and sensible environmental policies, could in fact lead to a better quality of environmental health.

A new era is creeping closer, one with perhaps zero harmful emissions, a new electric age for machinery and transportation. I look forward to riding my first electric tractor. This new technology, however, will be accompanied by a new way of looking at the world. This new worldview will carry with it a tacit understanding of coexistence, and a belief, although rarely acknowledged, that we have some control over the evolution of other species and our own.

Since we can monitor our environment like never before in Earth history, it should be our ethical imperative to try and maintain our species' gradual evolution by controlling the correlates of mass extinctions, namely carbon dioxide. This may be only one of many yet undiscovered environmental factors that lead to widespread extinctions, and we have to acknowledge our limitations. There's much that we cannot control—plate tectonics and its associated phenomena such as volcanoes and rifts come to mind. These processes cause greenhouse gases to be emitted into the atmosphere and oceans. Even though there's nothing we can do to prevent volcanic activity, however, inventors and futurists are working hard to develop innovations that can absorb or reuse greenhouse gases from all sources, even volcanic, in order to make our environment more sustainable.[206]

Instead of fighting conventional wars to eradicate "evil" people, pathogens, or ideologies, we can instead shift our ethical focus to managing the most fundamental factor in our evolution, the environment. By doing so we will also affect all other species of the biosphere. We can either accept the truth that the human population has already mushroomed to the point of affecting nearly all other species on the planet, or hide from it and pretend that we are an isolated population living a distinct and parallel existence with no need to care about others. If we accept the truth, we will have gone a long way toward adopting a new, promising worldview that we have some control over the future of population wars.

ACKNOWLEDGMENTS

I've been talking about this project for quite some time with friends, family, fans, and even journalists—basically anyone who asked, "Hey, Greg, what's your next book about?" That's the thing about books: They're easy to talk about but much harder to write. Without the constant support and encouragement of my agent and friend, Marc Gerald, this book would still be in the talking phase of development. He found the perfect editor to work with me, Peter Joseph at St. Martin's Press. I thank Peter for recognizing the potential of my idea for this book, and for making such excellent suggestions on the content, style, and overall vision of the final product. Peter's superb editorial skills improved the book immeasurably. I also benefited greatly from my friend and cowriter of the first draft, Caroline Greeven. She stitched together the initial chapters and helped guide the tone of the narrative with great craftsmanship.

It's a special privilege to have colleagues in academia, particularly those who specialize in evolution. I thank all of those in the Department of Ecology and Evolutionary Biology at Cornell, particularly Amy McCune, Nelson Hairston, and Rick Harrison. Rick and I have been teaching evolution together for the last five fall semesters. I've learned a lot about teaching from him, and benefited from the process. William Provine is an inspiration and good friend, from whom I continue to learn much. Also at Cornell, I'd like to thank Jon Parmenter of the History Department for suggesting a reading list on important Iroquois works.

Back in California, my academic colleagues are more like pals, since we went through undergraduate training together many years ago but still keep in touch. Thanks to Jay Phelan at UCLA, who wrote the best biology textbook for nonmajors, *What Is Life?*, Mark Gold of the UCLA Institute of the Environment and Sustainability, and Fritz Hertel, biology professor at Cal. State Northridge. Even though we didn't go to school together, Paul Abramson of the UCLA Department of Psychology, and the science writer Steve Olson, are good friends whom I can always count on for fun and writing advice.

I have the best family anyone could ask for. Allison, Stanley, Ella, Graham, Melikt, Amanuel, Grant and Lisa, and my wonderful in-laws, Frank and Sheila Kleinheinz—Mom, Dad and Julie, my love and appreciation goes out to you all, always and forever. My professional music partners are also some of my best friends. Thanks go to all of them for giving me the elbow room to accomplish this project alongside the various musical endeavors that constantly come our way: Jay Bentley, Brian Baker, Brooks Wackerman, Greg Hetson, Mike Dimkitch, Steven Barlevi, Frank Nuti, Eric Greenspan, Darryl Eaton, Ron Kimball, Cathy Mason, Tess Herrera. Special thanks go to my songwriting partner and pal since high school, Brett Gurewitz, for unflagging support in all things musical and intellectual. I also thank my out-of-the-limelight friends who helped in various ways: Don Ruff, Jeff Walden, John Lucas, and Paul Terry, for building my library and house; Bob and Mindi Fitzsimmons for being local sages of Schuyler County; Megan Shull for leading by example on how to be a prolific writer; Lori Perry for holding down the noise to a low "Sonic" boom while I wrote; Wryebo Martin and David Bragger, for always caring and engaging in fun activities, intellectual conversations, and creativity since we were kids; Mike Hove and Abbie Webb for the great companionship and conversations from neuroscience to environmental engineering (and plenty in between); and Chapter House Hockey for keeping competition constantly in the forefront of my thinking. I also want to thank all those behind the scenes who helped make this project successful: Melanie Fried, Sue Llewellyn, Laura Clark, Joanie Martinez, Christy D'Agostini, and everyone else at St. Martin's, and Karl Hensel at Kings Road Merch.

NOTES

INTRODUCTION: FINDING THE ENEMY

1. See *Sydney Morning Herald*, April 4, 2014, "War-Weary America Loses Appetite for Battle," p. 1, http://www.smh.com.au/world/warweary-america-loses-appetite-for-battle-20140404-zqql2.html.
2. A panmictic species is one in which any adult member of one population can successfully reproduce with an adult from another population within that species.
3. Those who blame and try to vilify the individual—Thomas Eric Duncan, the first person to die of Ebola on American soil—conveniently forget that he got the disease by heroically helping a pregnant Ebola victim, Marthalene Williams, in Africa who was too weak to walk. After carrying Ms. Williams back to her house bed, having been turned away from a Liberian hospital overrun by patients from the disease, Mr. Duncan likely was infected by contact with her. Ms. Williams died the next morning.
4. In fact there is a Bad Religion song on this topic called "Them and Us." on the album *The Gray Race*, Polypterus Music, BMI, Atlantic Recording Corporation (1996). Other songs mentioned in this paragraph: "California Uber Alles" (1979), single by the Dead Kennedys, Optional Music; "No Values" (1980) by Black Flag, *Jealous Again* ep, SST Records; "Wrecking Crew" (1982) by Adolescents, self-titled album, Frontier Records; "Fuck Authority" (1981) by Wasted Youth, *Reagan's In*, ICI Productions; "I Don't Care About You" (1982) by Fear, *The Record*, Slash Records.

CHAPTER ONE: PERSISTENCE IN THE FACE OF EXTINCTION

5. Loewen, James W. (2013), *Lies My Teacher Told Me: Everything Your American History Textbook Got Wrong*, Introduction, New Press, New York.
6. Preston, David L. (2009), *The Texture of Contact—European and Indian Settler*

Communities on the Frontiers of Iroquoia, 1667–1783, University of Nebraska Press, Lincoln, NE, p. 52.

7. Jordan, Kurt A. (2008), *The Seneca Restoration, 1715–1754: An Iroquois Local Political Economy*, University Press of Florida, Gainsville.
8. Cornell made his fortune in telegraph wire with his partner Samuel F. B. Morse, helped to build the first telegraph line in the United States (from Washington D.C. to Baltimore), and later invented the glass insulators that kept elevated wires free from short circuits where the wires met the poles. When Western Union was formed in 1856, Cornell's New York–to–Erie line was incorporated, and he earned millions in shares of the new company. This was the endowment used to start Cornell University.
9. In fact there is a town nearby called Podunk. I'm not sure if this is the geographical origin of the slang, but "Podunk town" has come to be a derogatory term for cities, whose citizens see small-town life as boring, pointless, and lonely.
10. The Pleistocene epoch is a division of Earth history that lasted from roughly 2.6 million years ago to about 10,000 years ago.
11. Allis-Chalmers tractors are legendary for their excellent design, functional utility, reliability, and simplicity of repair. Many small farms still depend on Allis Chalmers tractors today. My favorite machine is a 1962 Allis Chalmers D19 tractor that looks as cool today as it did when it was new. I use it as my "field" tractor, to cut and maintain the three-acre alfalfa field surrounding our house.
12. Even that other megalith of American manufacturing in Edison, New Jersey, General Electric, ordered parts from Allis-Chalmers. A seventy-five-ton generator rotor was machined by highly skilled laborers in Milwaukee to be installed at Consolidated Edison's Astoria, New York, plant in the 1950s. See Wendel, C. H. (2004), *The Allis-Chalmers Story*, Krause Publications, Iola, Wi, p. 119.
13. See the visitor statistics online here: http://newsdesk.si.edu/about/stats.

CHAPTER TWO: THE LONG HISTORY OF POPULATION WARS

14. In a biological sense the short-term goal might be limitless reproduction of offspring, or unlimited food resources. In a social sense it might be unlimited cash reserves, unlimited sexual partners, or total dictatorial power.
15. Much of the theoretical foundation work for predator-prey oscillations in population size was done by Alfred J. Lotka of Cornell and the University of Birmingham, and Vito Volterra at the University of Turin, in the 1920s. Although they never published together, both men are credited with applying statistical analysis to the relationships of predators and prey populations. Usually the field refers to their work as Lotka-Volterra equations. Many challenges to the accuracy of their equations, and to the assumptions built into them, have emerged over the years. Nonetheless Lotka-Volterra models still are taught as the foundation upon which predator and prey cycles can best be understood. See Ricklefs, Robert T., and Gary L. Miller (2000) *Ecology,* 4th ed., W.H. Freeman, New York, pp. 450–78. For the original studies on predator-prey populations in the laboratory, see Gause, G. F. (1934), *The Struggle for Existence*, Williams and Wilkins, Baltimore. Gause also coined the phrase "competitive exclusion," based on studies in the laboratory with *Saccharomyces* yeast: Gause, G. F. (1932), Experimental studies on

the struggle for existence: I. Mixed Population of Two Species of Yeast, *Journal of Experimental Biology* 9, 389. University of Birmingham (UK).

16. In fact it falls in at number three in my top ten list of life's greatest nuisances: (*1*) traffic, (*2*) bills, (*3*) plastic and cardboard prep. for recycling, (*4*) taking out the trash, (*5*) packaging, (*6*) software updates, (*7*) poorly mastered recordings, (*8*) poorly mastered television broadcasts, (*9*) political talk radio, (*10*) alert notification sounds on cell phones.
17. See White, Mathew (2013), *Atrocities: The 100 Deadliest Episodes in Human History*, Norton, New York. Some of history's worst wars really seem more like pogroms. I know this is a contentious issue, and I don't want to diminish the emotional impact that past injustices have on many people. But the fact that some readers might identify with being the "conquered" people merely underscores the fact that their group was not entirely "vanquished." Presumably there is still plenty of evidence that their ethnic heritage lives on in the communities and/or nations with which they identify. Most of what are considered "conquests" resulted, over the long term, in changes to human culture that eventually brought us to the present-day mosaic of global ethnicities. In short, the battles of yore have produced a mélange of cultural identities. The process by which this occurred was often completely unethical by today's standards, and often atrocious, as White's book attests, but the point I'm addressing in this book is not an ethical judgment but rather the long-term result of historical circumstances.
18. This is the age of some chemical "signatures" in rocks—that is, iron residues that show signs of rusting, which indicates the presence of oxygen from photosynthesis, a biological chemical reaction. But the first cyanobacteria-like colonial organisms have been dated from fossil evidence to roughly 3.5 bya. Remember that fossilization is a rare event, so fossils can only be used to set a "minimum" date; in other words the organisms almost certainly existed earlier but weren't fossilized.
19. And the suite of enzymes used to absorb free oxygen radicals (such as SOD, or superoxygen dismutase).

CHAPTER THREE: THE MEANING OF COEXISTENCE

20. This, of course, is only because we are comfortable. I realize that this statement would not apply if my family were in a position of squalor. In such families, improvement in their immediate needs is their primary concern.
21. It also must be acknowledged, however, that generalizing about the have-nots can border on sentimentalism and "bleeding-heart" assumptions that might not be true. For instance, I know that some homeless people are young runaways who are immaturely testing the limits of their own autonomy from overbearing parents. This rebellion is not the same as bona fide down-and-outs who've been dealt a brutal hand and genuinely need government assistance. We should not allow less needy youngsters "on the street" to sway our opinion that more agencies and facilities need to be built to house and care for the indigents and mentally ill who are homeless.
22. This term first appeared as the title of Nobel laureate Jacques Monod's book in 1970. In it he addresses some of the philosophical implications of evolution as a

purely mechanistic process. See also Provine, W. B. (1989), "Evolution and the Foundation of Ethics," in *Science, Technology, and Social Progress*, edited by Goldman, S. L., Lehigh University Press, Bethlehem, PA.

23. Recidivism, the relapse of criminals into further criminal behavior after intervention or sanctions (imprisonment), is a telling statistic. One study tracked nearly half a million convicted criminals who were punished as prisoners in 2005. Within three years of their release, 67 percent had returned to criminal activity and were rearrested. Within five years 76 percent were rearrested. If punishment were successful, we should expect to see lower recidivism rates and consequently a higher rate of rehabilitation. See http://www.nij.gov/topics/corrections/recidivism/Pages/welcome.aspx.
24. Regardless of the complicated causes of each, the Bloody Christmas of 1951 and the Watts Riots of 1965—both violent clashes between LA's minority communities and the LAPD—were still fresh in most African-Americans' memories in 1994. More recently the killing of an unarmed teenager, Trayvon Martin, in Florida, and the exoneration of his killer, George Zimmerman, was still fresh in the memories of the outraged citizens of Ferguson, Missouri, when police killed Michael Brown, also an unarmed teenager.
25. It's important to distinguish sickle-cell trait (SCT) from sickle-cell anemia (SCA). SCT is found in carriers who have inherited the sickle-cell gene variant (allele) from only one parent. The red blood cells are sickle shaped, which provides malarial resistance, but SCT does not cause anemia. SCT individuals are considered heterozygous for the sickle-cell allele. SCA is more severe because carriers have received the sickle-cell gene variant from both parents. SCA individuals are considered homozygous for the sickle-cell allele.
26. *Supreme Court Justice Stevens, January 2010,* "We the People, Not We the Corporations," www.movetoamend.org.

CHAPTER FOUR: THE CONTEXT OF PERSISTENCE, THE BACTERIAL DIMENSION

27. See Nelson and Cox (2005), *Lehninger's Principles of Biochemistry*, 4th edition, p. 252.
28. Some oil shales come from accumulations of plant debris, particularly from the Pennsylvanian and Carboniferous swamps.
29. Remember, a bacterial generation might last only thirty minutes or less.
30. This is why so many municipalities treat "city" water with chlorine, a chemical additive that kills anaerobic bacteria.
31. See Dobell, Clifford (1932), *Antony van Leeuwenhoek and his "Little Animals",* Harcourt, Brace and Co., New York.
32. See Nelson and Cox (2005), *Leninger's Principles of Biochemistry,* p. 252.
33. http://www.cdc.gov/drugresistance/threat-report-2013/.
34. See http://www.usatoday.com/story/news/nation/2013/05/22/portland-fluoride-water/2350329/.
35. Fermentation acids of both *Streptococcus mutans*, *Streptococcus sobrinus*, and some *Lactobacilli* cause dental caries, aka cavities.
36. Sometimes called blue-green algae.
37. Using extremely sophisticated probes, scientists can measure isotopic variation in

ancient sediments—particularly useful for determining photosynthesis is carbon. The slight differences in ratios of carbon-13 and carbon-12, for instance, reveal the history of biological activity during the deposition of the sediment. Photosynthetic organisms prefer to use the lighter form of carbon (^{12}C) resulting in an enrichment of heavy ^{13}C in the sediments. Similar isotopic ratio analyses (also called fractionation experiments) of elements such as sulfur and oxygen reveal a variety of ancient environmental clues. Oxygen isotopes can reveal the relative amounts of glacial ice vs. liquid water on the planet (ice tends to bind the lighter oxygen isotope leaving the heavier isotope available for plants to use as CO_2, and animals to use for their biological needs). Sulfur isotopes can be used to determine the presence of sulfate SO_4 in the early ocean, which, as mentioned in the text, can be an indication of pyrite precipitation and hydrogen sulfide occurrence. See Knoll, Andrew H. (2003), *Life on a Young Planet,* Princeton University Press, p. 102.

38. This is an accepted number among many microbiologists. See Lee, Y. K. (2008), *Who Are We? Microbes the Puppet Masters,* World Scientific, Singapore, p. 29.
39. Species is a concept that is not easily applied to bacteria. Although we conventionally describe them with binomial nomenclature, such as *Helicobacter pylori*, microbiologists such as Y. K. Lee (2008) prefer to use the term "strain" to describe different genetic variants of bacteria.
40. See Walter, Jens and Ruth Ley (2011), The human gut microbiome: ecology and recent evolutionary changes, *Annual Review of Microbiology* 65, 411–29.
41. http://www.scientificamerican.com/article/the-guts-microbiome-changes-diet/
42. Ibid., p. 413.
43. Roughly 10 percent of our daily nutrition comes from the fermentation products of bacteria in our guts. See Bersaglieri, T., P. Sabeti, N. Patterson, T. Vanderploeg, S. Schaffner et al. (2004), Genetic signatures of strong recent positive selection at the lactase gene, *Am. J. Hum. Genetics* 74, no. 6, 1111.
44. Walter and Ley (2011), p. 412.
45. See Cryan, John F. and Timothy G. Dinan (2012), Mind-altering microorganisms: the impact of the gut microbiota on brain and behavior, *Nature Neuroscience* 13, 701.
46. See David, Lawrence A., et al. (2014), Diet rapidly and reproducibly alters the human gut microbiome, *Nature* 505, 559.
47. Certain bacteria live in yogurt that we eat (sometimes called "probiotics" or "live cultures"). These species enter our guts where they continue to live and reproduce. In the gut they digest sugars from dairy products (lactose) by secreting an enzyme (lactase). The production of lactase benefits us in digestion, and recent studies have demonstrated that these bacteria are associated with lowering the prevalence of cancerous tumors.
48. Lee, Y. K. (2008), *Who Are We? Microbes the Puppet Masters*, World Scientific, Singapore.
49. Ibid., p. 25.
50. http://www.historytoday.com/mary-harlow/old-age-ancient-rome.
51. Lee, Y. K. (2008), *Who Are We? Microbes the Puppet Masters*, p. 25.
52. See Rohwer, Forest (2003), Global phage diversity, *Cell* 113, no. 2, 141; Mora, Camilo, et al. (2011), How many species are there on earth and in the ocean? *PLoS*

Biol 9, no. 8; and Doolittle, W. Ford, and O. Zhaxybayeva (2009), On the origin of prokaryotic species, *Genome Research* 19, no. 5, 744.

CHAPTER FIVE: THE SYMBIOTIC DEPENDENCY OF LIFE, THE VIRAL DIMENSION

53. Specifically, the liver receives blood from the capillaries of the stomach, spleen, pancreas, and hepatic artery, passes this blood through a sinusoidal network of microscopic lobules containing hepatocytes (aka liver cells) that are active in metabolizing food, producing necessary building blocks such as proteins, glucose (glycogen), fatty acids, and cholesterol, filtering exogenous toxins, storing antibodies, eating excess bacteria from the gut, storing blood, and producing bile for digestion, among other functions. The products of the liver are released into the bloodstream, from which they return to the heart to be recirculated through the arterial system.
54. Although some viruses can be found in the sea or other watery habitats, they exist there in a state of "suspended animation" or "inert potential," not replicating nor doing much at all metabolically. Rather they lie in wait for other passers-by to parasitize.
55. This is called lysogenic integration, whereby the replication of the viral parasite is inactivated. The more familiar kind of viral infection, as in the common cold or flu, is called lytic integration, in which case the replicative machinery is kicked into high gear and numerous copies of daughter viruses are produced. Lysogenic integration is often characterized by retroviruses that use RNA rather than DNA as their genetic material.
56. See Ting, C.-N., et al. (1992), Endogenous retroviral sequences are required for tissue-specific expression of a human salivary amylase gene, *Genes & Development* 6, no. 8, 1457.
57. See Weiss, R. A., and J. P. Stoye, (2013), Our viral inheritance, *Science* 340, no. 6134, 820.
58. Of course this should not be taken to mean that your life is irrelevant. Human meaning is transmitted by culture, so your contributions might be highly significant even if your genetic heritage is not passed on to future generations.
59. For DNA these letters are AGTC, for adenine, guanine, thymine, and cystosine. For RNA, uracil (U) is substituted for thymine. DNA and RNA are nucleic acids. The other letters represent nucleobases.
60. Retroviruses use RNA as their genetic material and produce an enzyme called reverse transcriptase to produce DNA once inside the host cell. The newly formed DNA is then inserted into the host's own DNA, where it rests. At this resting state it is sometimes called a provirus.
61. See Ryan, Frank (2009), *Virolution*, 116–17, Collins, London.
62. *Homo habilis* is the oldest humanlike fossil that belongs to our genus. It is known from deposits in Africa that are roughly 2.3 million years old.
63. Sentimentality, in my opinion, should be relegated to its very important role, as a family "glue" or a means of enhancing social connections.
64. The origin of eukaryotic cells is important to note here. Bacterial endogenization occurred sometime before 1.6 bya when a free-living bacterium took up residence inside a free-living photosynthetic algae. This endosymbiosis was highly selec-

tively advantageous, and it resulted in successful reproduction and massive buildup of cells and colonies of eukaryotic organisms. The bacterial portion of the eukaryotic cell provides engine-like services to the host cell. ATP for instance, which is like a molecular fuel for energy, comes from organelles called mitochondria that have their own DNA separate from that of the host cell's nucleus. This DNA matches that of some types of free-living bacteria called proteobacteria. This matching of DNA is the strongest evidence for the eukaryotic cell's endosymbiotic origin. See Margulis, Lynn (1996), Archaeal-eubacterial mergers in the origin of Eukarya: Phylogenetic classification of life, *Proceedings of the U.S. National Academy of Sciences* 93, no. 3, 1071.

65. See Margulis, Lynn, et al. (2000) The chimeric eukaryote: Origin of the nucleus from the karyomastigont in amitochondriate protists, *Proceedings of the National Academy of Sciences* 97, no. 13, 6954.
66. A virus is essentially a collection of genetic material (DNA or RNA) wrapped in a protein coat, or capsid. Most capsids are enclosed inside an outer membrane, or envelope, composed of glycoprotein and lipid. Some viruses known as bacteriophages specialize on infecting bacteria and have rudimentary morphology. They have a head—the capsid—and a neck, and a tail sheathed in proteins with small "fibers" at the most distal projection. Other viruses lack the component "body" parts and are simply spheroid masses with fibrous projections coming off in multiple directions. These fibers contain the glycoproteins necessary to attach to specific host-cell surfaces.
67. For purposes of explanatory simplicity, I use DNA as an example of genetic material. Please be aware, however, that many viruses use RNA, a slightly different molecule, as their genetic material.
68. See Gupta, Radhey, and G. Brian Golding, (1996), The origin of the eukarotic cell, *Trends in Biochemical Sciences* 21, no. 5, 166.
69. See Gupta, Radhey (1998), Protein phylogenies and signature sequences: A reappraisal of evolutionary relationships among Archaebacteria, Eubacteria, and Eukaryotes, *Microbiology and Molecular Biology Reviews* 62, no. 4, 1435.
70. Interestingly, it is hypothesized that the plastids were once free-living cyanobacteria. The origin of plants, therefore, according to this view, was the endosymbiotic union of one of the first photosynthetic organisms (cyanobacteria) with an already-established eukaryote sometime in the distant Precambrian. See Gupta (1998), figure 1.
71. The best summary of viral endosymbiotic interactions is found in Villarreal, Luis P., and Guenthea Witzany (2010), Viruses are essential agents within the roots and stem of the tree of life, *Journal of Theoretical Biology* 262, no. 4, 698.
72. By "familiar" I mean to draw attention toward things like polar bears, dogs, cats, humans, mice, corn, trees, flowers, amoebas, parameciums, and so on. Although it's a bit nebulous—what's familiar to one person might not be familiar to another—it is preferable to that taxonomic awkwardness of calling eukaryotic organisms "higher" and prokaryotic organisms "lower" life forms, which flies in the face of the argument in this chapter—that is, evolution didn't appear on this planet as a Scala Naturae. If need be the reader can equate "familiar" with the term "higher organisms." I am assuming no one is really familiar with bacteria (prokaryotes) except specialists, so they could be thought of as "lower," but only in the colloquial sense.

73. Viruses mutate exceptionally fast. Not by substitutions of single nucleic acids (aka point mutations) but rather by wholesale excising and replacement of entire genes, that is, transposition. It is known that viral resistance to HIV drugs can evolve within a strain as quickly as one day after exposure to the drug. See Sanjuan, Rafael, et al. (2010) Viral mutation rates, *Journal of Virology* 84, no. 19, 9733. It should also be noted that transposons (jumping genes) were discovered in corn by Barbara McClintock of Cornell University in 1931. For a lifetime of work in plant breeding she was finally awarded the Nobel Prize for Physiology or Medicine in 1983.
74. And there is more supporting evidence: For instance virus DNA is eukaryote-like in the sense that it is linear strands of genetic material as opposed to the ringlike DNA of prokaryotes. This linear arrangement requires a different—some would say more elaborate—set of replication enzymes and proteins than is found in prokaryotes. There are more starts and stops, for instance, when replicating linear segments of multiple chromosomes (eukaryotic and viral) as opposed to a single start and stop in a circular strand (prokaryotic). And finally, the free-floating mimiviruses, large, membrane-bound DNA viruses, are common in the oceans today. They have many properties that make them good candidates as descendants of the ancestral organism that infected another free-floating cell in the Precambrian, forming an endosymbiotic union and functioning as the primordial nucleus. See Bell, P. J. L., (2001), Viral Eukaryogenesis: Was the ancestor of the nucleus a complex DNA virus?, *Journal of Molecular Evolution* 53, no. 3, 251, and Monier, Adam, Jean Michael Clavier, and Hioyuki Ogata (2008), Taxonomic distribution of large DNA viruses in the sea, *Genome Biology* 9, no. 7, R106.
75. The human genome contains an abundance of HERVs. HERV stands for human endogenous retrovirus." The size of the human genetic load that is made up of HERVs is surprising in many ways. HERVs make up roughly 8 percent of the total human genome. But when we add other components of our genome that are direct viral descendants or viral dependent for their replication, the proportion jumps to nearly half of the human genome. This includes LINEs, an acronym for "long interspersed nuclear elements (portions of the genome that code for an enzyme called reverse transcriptase, an essential viral enzyme thought to have made the jump to vertebrates and other organisms in the distant evolutionary past), and SINEs, short interspersed nuclear elements (portions of the genome that are not technically of viral origin, and "noncoding," meaning that they are not used for making proteins per se but may be crucial in modifying the activity of other proteins. So-called alu sequences are the most common type of SINE. Since they depend on HERVs and LINEs, they are considered retrotransposons, which are uniquely viral in function. See Cordaux, Richard, and M. Batzer (2008), The impact of retrotransposons on human genome evolution, *Nature Reviews Genetics* 10, 691; and Ryan, Frank (2009), *Virolution*, Collins, London, pp. 172–74.
76. It was Herbert Spencer who coined and promoted the phrase "survival of the fittest," not Charles Darwin. It was in the promotion of Darwinism, however, that Spencer's phrase was used so often. It has become nearly synonymous with "natural selection" and is therefore viewed as part of the Darwinian doctrine. See Spencer, Herbert (1895), *Principles of Biology*, Appleton & Co., New York, vol. 2, p. 478.
77. See Ryan, Frank (2009), *Virolution*, Collins, London.

78. They may do terrible damage when they find a new host species—for example, smallpox, when it reached a formerly uninfected population of Native Americans, wiped out up to 50 percent of those it infected. The same percentage is true of first contact of Native Americans and the plague (see Hunt, George T. [1960], *The Wars of the Iroquois,* Univ. of Wisconsin Press, p. 40). For those who weren't killed by smallpox or plague, there were still severe and damaging injuries that persisted for the rest of their lives due to the infection. But since viruses, like bacterial infections, do not kill the entire population, they either persist in low numbers (as smallpox is doing currently), become extinct, or hybridize with the host genome (as the endogenous retroviruses have done). If viruses succeed at the latter, they may participate in the evolution of the species.
79. It has been shown that undifferentiated and highly proliferative cell lines as seen in early embryos show high levels of the enzyme reverse transcriptase (RT). Terminal states of cells, as seen in late stages of differentiation, such as in adult somatic tissues, show low levels of RT. See Spadafora, Corrado (2008), A reverse transcriptase-dependent mechanism plays central roles in fundamental biological processes, *Systems Biology in Reproductive Medicine* 54, no. 1, 11.
80. In certain cancers that are caused by retroviruses such as some leukemias, anti reverse transcriptase drugs have shown promise. In other trials, HIV, also caused by a retrovirus, was impaired by application of anti-RT drugs.

CHAPTER SIX: ESTABLISHING A WAR NARRATIVE FOR POPULATIONS, THE IMMUNE SYSTEM

81. Many viruses carry only RNA as their genetic material. As in the last chapter, when I reference DNA I am using it as a shorthand for genetically replicating material and including RNA-based organisms in my statement. When I distinguish RNA from DNA it will be clear from the context of the text.
82. This analogy could equally apply to any of the higher taxonomic groups of organisms, such as families, orders, classes, phyla, or even kingdoms.
83. See Scott, G. R. (2011), Elevated performance: The unique physiology of birds that fly at high altitudes, *Journal of Experimental Biology* 214, 2455.
84. See, for instance, Sompayrac, Lauren (2012), *How the Immune System Works,* 4th ed., Wiley-Blackwell, Hoboken.
85. "Germ" and "pathogen" are used interchangeably to indicate a microbe that causes disease.
86. See Echenberg, Myron J. (2002), Pestis redux: The initial years of the third bubonic plague pandemic, *Journal of World History* 13, no. 2, 429.
87. *Yersinia pestis* can take on different pathologies in mammals. For instance, certain variants take up residence in lung tissue and cause pneumonic plague. These germs can be spread by coughing and sneezing, from human to human. Other variants of *Yersinia pestis* are restricted to the bloodstream and cause a disease called septicemia, or septicemic plague. The buboes, or painful swellings of the skin associated with bubonic plague, are due to massive population increase of *Yersinia pestis* in the lymph nodes.
88. See Samia, N. I., et al. (2011), Dynamics of the plague-wildlife-human system in central Asia are controlled by two epidemiological thresholds, *Proceedings of the National Academy of Sciences,* (USA) 106, no. 35, 14527–32.

89. See http://emedicine.medscape.com/article/829233-overview#a0199.
90. Phagocytosis is the cellular process of engulfing other cells, such as microbial pathogens, or toxic particles, or as we saw in the last chapter in the case of endosymbiosis. The engulfed cells or particles are rendered harmless through sequestration inside an organelle (phagosome) or through enzymatic digestion. It is also known as endocytosis. See Sun, W., et al. (2013), Pathogenicity of *Yersinia pestis* synthesis of 1-dephosphorylated lipid-A, *Infection and Immunity* 81, no. 4, 1172.
91. http://www.pbs.org/wgbh/pages/frontline/tb-silent-killer/.
92. See Sompayrac, Lauren (2012), 13. The boundary between the Precambrian and the Cambrian, or more specifically between the Proterozoic eon and the Paleozoic eon, is 540 million years ago. The Paleozoic (or "ancient life" eon) contains fossils of the oldest known vertebrates—our ancestors that had a "backbone," cartiliganous or bony skeleton, and teeth or denticles made of dentine. Before this date the fossil record is relatively poor. But some exceptional fossil localities have been found that contain probable ancestors of both vertebrates and the group to which sea urchins belong. For more see Fedonkin, M. A., et al. (2007), *The Rise of Animals, Evolution and Diversification of the Kingdom Animalia,* Johns Hopkins, Baltimore.
93. Also, sea urchin relatives, the sea stars (all members of the phylum Echinodermata), have organs that are analogous to the immune functions of the vertebrate liver. These organs in echinoderms produce complement proteins just as the vertebrate liver does. This indicates that the ability to produce complement has an ancient inheritance in animals. See Smith, L. C. Courtney, Lori A. Clow, and David P. Terwilliger. (2001), The ancestral complement system in sea urchins, *Immunological Reviews* 180, 16; and Leclerc, Michel, Nicolas Kresdorn, and Björn Rotter (2013), Evidence of complement genes in the sea-star *Asterias rubens*: Comparisons with the sea urchin, *Immunology Letters* 151, nos. 1–2, 68.
94. I jokingly tell my students, "When an evil in the world is met with compliments [complements], it quickly becomes benign." This is a nerdy way of saying "Flattery conquers all," despite the difference in spelling.
95. In this discussion I am lumping dendritic cells in with macrophages, even though some essential differences exist between the two. See Banchereau, Jacques (2002), The long arm of the immune system, *New Answers for Cancer, Scientific American Special Edition* 287, no. 5, 58.
96. Remember that prokaryotic cells (bacteria) and viruses are tens to hundreds of times smaller than the larger eukaryotic cells (such as amoebas). A marble and a medium-size balloon make good analogies in this illustration.
97. In fact, certain bacteria such as the pathogen that causes Legionnaire's disease, *Legionella sp.*, is able to infect amoebas and other pond-dwelling protists in addition to human macrophages. In both cases the bacterium uses the highly conserved phagocytic "machinery" of these cells to enter and reproduce itself inside the host's cytoplasm. The point to take home is that phagocytic activity has remained essentially unchanged since the common ancestor of amoebas and humans, prior to 2.1 bya. What has changed is the symbiotic associations of phagocytic cells over time. See Al-Quadan, Tasneem, Christopher T. Price, and Yousef Abu Kwait, (2012), Exploitation of evolutionarily conserved amoeba and mammalian processes by Legionella, *Trends in Microbiology* 20, no. 6, 299.

98. See Sun, W., et al. (2013), Pathogenicity of *Yersinia pestis* Synthesis of 1-Dephosphorylated Lipid A, *Infection and Immunity* 81, no. 4, 1172–85.
99. http://www.timeshighereducation.co.uk/198208.article.
100. NK (natural killer) cells are another type of dedicated foot soldier that has no antigen-presenting function. They can inject enzymes into pathogens, and these enzymes cause the microbes to "commit suicide." Other "soldiers" include eosinophils and mast cells that specifically target amoebas or parasites, and monocytes that leave the bloodstream to mature into macrophages that reside within the muscles, skin, or connective tissues.
101. The nineteenth-century German zoologist Ernst Haeckel popularized a truism in biology with his phrase, "Ontogeny recapitulates phylogeny." It had long been observed that embryos of mammals and birds go through a series of developmental stages (called ontogeny) that appear to resemble the adult stages of fishes, amphibians, and reptiles. Evolutionary theory, around the time of Darwin, led many investigators to conclude that embryos depict a sequence of evolutionary ancestors (phylogeny). Although adult stages of ancestors are clearly not represented in embryos of descendant species, modern evolutionary developmental genetics (aka EVO-DEVO) shows us that embryonic development (and therefore adult anatomy) is controlled by particular genes that were passed from ancestor to descendant over the course of evolutionary time. The case can be made, therefore, that development of particular tissues and organs is based on ancestral conditions and not created de novo. This is true for cell lineages as well, such as macrophages being derived from ancestors that possessed only monocytes. For more on ontogeny see Gould, Stephen J. (1974), *Ontogeny and Phylogeny*, Belknap, Harvard. For more on the evolutionary ancestry of immune cells see Boehm, Thomas (2012), Evolution of vertebrate immunity, *Current Biology* 22, no. 17, R722–85.
102. Lampreys, one of the groups of living jawless fishes, come from an ancient lineage that was ancestral to our own vertebrate stem group. Lampreys and their ilk share a common ancestor with us that is hypothesized to be roughly 530 million years old. See Dawkins, Richard (2004), *The Ancestor's Tale*, p. 354, Houghton Mifflin, New York. Lampreys not only have lymphocyte-like cells that react with common antigens, they are located in areas where "higher" forms, such as humans, have lymphoid organs. Hence it seems that vertebrate adaptive immunity was present in rudimentary form in our oldest ancestors. See Litman, Gary W. Jonathan P. Rast, and Sebastian D. Fugmann (2010), The origins of vertebrate adaptive immunity, *Nature Reviews Immunology* 10, no. 8, 543; and Boehm, Thomas (2012), Evolution of vertebrate immunity, *Current Biology* 22, no. 17, R722.
103. The process described here is a subset of T-cells known as "killer T-cells" (aka cytotoxic T-cells). The three main types of T-cells are "killer T-cells" that specialize in killing any cell that has been infected by a virus, "helper T-cells" that secrete chemical signals (cytokines, e.g. interleukin II, interferon gamma, and so on) to alert other cells that an infection is occurring. (Helper T-cells can be thought of as cytokine factories just as B-cells are antibody factories). And a third type of T-cell that is the least understood, "regulatory T-cells," which assist other components of the immune system by chemical signals that report on the progress of the infection.

104. Sompayrac, Lauren (2012), *How the Immune System Works,* 4th ed., Wiley-Blackwell, Hoboken, p. 5; Tonegawa, Susumu, et al. (1977), Dynamics of immunoglobulin genes, *Immunological Reviews* 36, 73.
105. In going from an immature B-cell to a mature B-cell, many rounds of cell division take place in the circulatory systems. The immature B-cell is different from its progeny in that the offspring have been subject to numerous rounds of genetic recombination and selective winnowing of the original parental variation. There is also a lot of cell death. Many unsuccessful genetic combinations take place in the assembly of antibodies, so only some of the variants go on to further rounds of cell division. With each round, the B-cell lineage matures and becomes committed to making only one type of antigen. These mature B-cells are also called plasma cells.
106. The technical term for "proper antibody" is B-cell receptor (BCR) or T-cell receptor (TCR). A receptor is a molecule that combines with another molecule (in this case an antigen molecule on the surface of a pathogen) and brings about a change in cell function (in the case of B-cells, when the receptor meets its paired antigen, it begins to make antibodies; in the case of killer T-cells, a matched pairing brings about the intimate contact and dismantling of the pathogen's cell membrane).
107. See Litman, Gary W., Jonathan P. Rast, and Sebastian D. Fugmann (2013) The origins of adaptive immunity, *Nature Reviews Immunology* 10, no. 8, 543.
108. In actuality the innate and adaptive immune systems work in close collaboration during an infection, but for this discussion I am focusing on the adaptive immune response for simplicity. See Murphy, Kenneth, and Paul Travers (2012), *Janeway's Immunobiology*, 8th ed., Garland Science, New York. Chap. 16 and appendix 1.
109. Tonegawa, Susumu, et al. (1977) Dynamics of immunoglobulin genes, *Immunological Reviews*, 36, 73; pg 91.

CHAPTER SEVEN: WAR IS UNWINNABLE

110. Huxley, Thomas H. (1894), *Evolution and Ethics*, Prolegomena, Appleton Press, New York, p. 8.
111. See Fischer, David Hackett (2009), *Champlain's Dream: The European Founding of North America*, Simon & Schuster, New York; an excellent map showing the Indian nations and their trade routes can be found on p. 135.
112. The area sometimes called Iroquoia by historians has had different geographical boundaries at various stages of history, depending on the population size and political arrangements with neighboring Indian nations or with Europeans. Freedom of movement for hunting and war parties also affected the outline of their territory on historical maps. But at various times Iroquoia stretched roughly from Montreal, Quebec, nearly to Cleveland along the southern edges of Lake Ontario, to Lake Erie. Iroquois controlled the upper drainages of the Ohio River as well as the Susquehanna and Hudson Rivers. This latter drainage marked the eastward terminus of their territory. In the middle of this vast region sit the Finger Lakes of New York, and the Iroquois "capital," where all important councils were held, at Onondaga, a town bordering the lake of the same name near present-day Syracuse.

    It was here, at Onondaga, that sachems would congregate from all over

Iroquoia to deliberate important issues. Usually fifty in number, these leaders represented widely spaced localities, and gave beautifully orated speeches that were debated over a period of days. The end of a council was marked by a consensus, and the sachems returned to their village or town and disseminated the terms. In this way the Iroquois Confederacy, made up of five nations (Mohawk, Oneida, Onondaga, Cayuga, and Seneca—and later a sixth added, the Tuscarora), was held together as a political unit. See Norton, A. Tiffany (1879), *History of Sullivan's Campaign Against the Iroquois, Being a Full Acccounting of that Epoch of the Revolution*, Norton, Lima, N.Y., p. 10, and Parmenter, Jon (2010), *The Edge of the Woods: Iroquoia 1534–1701*, Michigan State University Press, p. xiii.

113. Hunt George T. (1960), *The Wars of the Iroquois: A Study in Intertribal Trade Relations*, University of Wisconsin Press, p. 28.
114. Ibid., p. 52.
115. Ibid., p. 70. The Dutch would not hesitate to trade guns with the Iroquois. But the French generally refused to trade guns with Indians. By 1641 the most striking fact of Iroquois trade was that they only had thirty-six muskets (arquebuses) in their band of five hundred traders who came to Montreal with a French captive with hopes of trading him for more guns. The French governor Montmagny, insisted instead on peaceful relations with the Hurons, and he gave no guns. This was the beginning of the Iroquois wars, driven by economic need, according to Hunt.
116. Ibid., p. 82–83.
117. See Greer, Alan (2000), *The Jesuit Relations: Natives and Missionaries in Seventeenth-Century North America,* Bedford/St.Martins, Boston, p. 111.
118. Fischer, David Hackett (2009) *Champlain's Dream*, p. 7. ". . . [Champlain] envisioned a new world as a place where people of different cultures could live together in amity and accord. This became his grand design for North America."
119. See Preston, David L. (2009) *The Texture of Contact: European and Indian Settler Communities on the Frontiers of Iroquoia, 1667–1783,* University of Nebraska Press, Lincoln. The "shade from the tree of peace" appears on p. 26.
120. Ibid. *Haudenosaunee* is the native Iroquoian word for "people of the longhouse." It is used synonymously by historians to denote the five (or six) nations of the Iroquois Confederacy.
121. As evidenced by narrative tradition, this ancient ritual of adopting enemies goes back hundreds of years in Iroquois culture. The five nations warred constantly before contact with Europeans, as the epic myth states, and this produced great mourning and sadness. In order to cope with the death of loved ones a "mourning war" ritual became the norm. "At the request of female kin of a deceased person, warriors raided a traditional enemy for captives, who would then face one of two fates. Either they would be adopted by a grieving family as an almost literal replacement for the departed, or they would be executed in rituals." From Richter, Daniel K. (1987), "The Ordeal of the Longhouse: The Five Nations in Early American History," in *Beyond the Covenant Chain, The Iroquois and Their Neighbors in Indian North America*, 1600–1800, Richter, Daniel K. and J. H. Merrell, eds. Pennsylvania State University Press, University Park.
122. As an example, Onondagas had a plan for their wars: ". . . to adopt prisoners and captives; that fragments of their tribes were parted amongst them, and thus lost.

They used the term *We-hait-wa-tsha,* in a figurative sense in relation to such tribes. This term means a body cut and quartered and scattered around. So they aimed to scatter their prisoners among the other nations. There is still blood of the Cherokees in Onondaga." Schoolcraft, Henry Rowe (1847), *Notes on the Iroquois, or Contributions to American History and Antiquities, and General Ethnology,* Erastus Pease & Co., Albany, NY p. 443.

123. Hunt, George T. (1960), "The Hurons and Their Neighbors," p. 116.
124. "If we understand colonization to represent the conquest of a physical space by an alien polity and that polity's ability to set the terms by which conquered space is defined, one cannot refer to Iroquoia as colonized in 1701." Parmenter, Jon (2010), *The Edge of the Woods*, Michigan State University Press, Lansing, xxxiv.
125. From Norton, A. Tiffany (1879), *History of Sullivan's Campaign Against the Iroquois, Being a Full Acccounting of that Epoch of the Revolution*, A. T. Norton, Lima, NY.
126. See Abler, Thomas S. (2007), *Cornplanter, Chief Warrior of the Allegany Senecas*, Syracuse University Press, p. 15.
127. In fact not all members of the confederacy agreed to become allies of the British. In the long-standing spirit of the councils at Onondaga, the Iroquois sachems usually put into practice only those agreements that were unanimous. If unanimity could not be achieved, each dissenting group could take an autonomous course. In the case of the great council on entering the Revolutionary War, the sachems of the Oneida nation pledged neutrality. Warriors from the other Iroquois nations—Cayuga, Seneca, Mohawk, and Onondaga—joined forces with the British. William Johnson died on the eve of the Revolutionary War, in 1774. He stated that he would never want to live to see the day when his Iroquois friends should ever have to engage in a war for or against Britain (see Norton, *History of Sullivan's Campaign*, pp. 28–29).
128. ". . . . the King [of England] they declared to be rich and powerful, his rum was as plenty as the water in Lake Ontario, and his men as numerous as the sands upon the lake shore. If the Indians would lend their assistance [in the Revolutionary War] they should never want for money or goods. This appeal to their avarice overcame the scruples of the Indians, and with the exception of the larger portion of the Oneidas, they concluded a treaty with British agents in which they engaged to take up arms against the Colonists, and continue in the King's service until they were subdued. From Norton, A. Tiffany (1879), *History of Sullivan's Campaign Against the Iroquois, Being a Full Acccounting of that Epoch of the Revolution*, Norton, Lima, NY, p. 50.
129. See Fischer, J. R. (1997), *A Well-Executed Failure: The Sullivan Campaign Against the Iroquois, July–September 1779*, University of South Carolina Press, Columbia, pp. 25–30.
130. Ibid.
131. The previously mentioned Battle of Oriskany was part of this effort. Formerly British, Fort Stanwix was built in 1758 by the British to end French incursions into the Mohawk Valley during the French and Indian War. Near the end of that war, in 1760, the fort was abandoned. By 1777 George Washington ordered Maj. Gen. Philip Schuyler, who presided over the northern army in the Revolutionary War, to have it refurbished.

132. See, for example, Turner, O. (1851), *History of the Pioneer Settlement of Phelps and Gorham's Purchase and Morris' Reserve; Embracing the Counties of Monroe, Ontario, Livingstone, Yates, Steuben, Most of Wayne and Allegany, and Parts of Orleans, Genesee, and Wyoming, to which is Added, A Supplement, or Extension of the Pioneer History of Monroe County*, William Alling, Rochester, NY.
133. Norton, A. Tiffany (1879), *History of Sullivan's Campaign*, p. 67. The use of the word "savages" has had a long history. It began as a translation into English from the French, which described the North American Indians as *sauvages*, or "people of the woods." Obviously the translation has become an unfortunate derogation.
134. Ibid., pp. 75–80, Sullivan's letter to Washington, detailing his understanding of the purpose for the expedition.
135. Ibid., Washington's instructions to Sullivan.
136. Lt. William Barton's journal: "Monday August 30th At the request of Maj. Piatt, sent out a small party to look for some of the dead Indians—returned without finding them. Toward noon they found them and skinned two of them from their hips down for boot legs; one pair for the Major the other for myself. . . . Tuesday, Sept. 14th . . . At this place was Lieut. Boyd and one soldier found, with their heads cut off; the Lieut'nts head lay near his body; his body appeared to have been whipped and pierced in many different places. The others head was not found. A great part of his body was skinned leaving the ribs bare." Cook, Frederick (1885), *Journals of the Military Expedition of Major General John Sullivan Against the Six Nations of Indians in 1779*, Books for Libraries Press, Freeport, NY; pp. 8, 11.
137. See Parmenter, Jon (1999), "Isabel Montour: Cultural Broker on the Frontiers of New York and Pennsylvania," in *The Human Tradition in Colonial America*, edited by Ian K. Steele and Nancy L. Rhoden, Wilmington, DE.
138. This is the view of A. Tiffany Norton (1879), *History of Sullivan's Campaign*, pp. 120–21: "Had the [battle of Newtown]" resulted in defeat to our arms, it is difficult to measure the degree of misfortune that would have followed to the cause of the Colonies. . . . the enemy would probably have followed up its success until the Colonial army had been driven back to its starting point, and elated by victory as well as inspired by a desire for revenge, the savage foe would have wreaked terrible vengeance on the unprotected frontiers. . . . The utter defeat of the chief military movement of that year might have so disheartened the Colonies, at a period when disaster was doubly dangerous, that they would have given the further prosecution of the war a feeble support or abandoned it altogether; while the British government, inspired by such success, would have put new vigor into its measures for the suppression of the revolution. Without weighing, therefore, the influence which Sullivan's campaign subsequently had in directing the march of civilization toward the rich country of Western New York, the Battle of Newtown may justly be considered one of the most important engagements of the Revolutionary War, and as worthy of commemoration as Bunker Hill or Monmouth, Brandywine or Princeton." Others challenge this view and see the Sullivan campaign as a political and humanitarian disaster at this stage of the Revolutionary War. See Fischer, J. (1997), *A Well-Executed Failure, The Sullivan Campaign Against the Iroquois, July–September 1779*, University of South Carolina, Columbia.

139. Washington's general orders to the army, Oct. 17, 1779, Norton A. Tiffany, *History of Sullivan's Campaign*, (1879), p. 180.
140. Sullivan's transmission to Congress, November, 1779, Ibid., p. 182.
141. Kayangaraghanta to Guy Johnson, Dec. 16, 1779, Fischer, J. R. (1997), *A Well-Executed Failure, the Sullivan Campaign Against the Iroquois, July-September 1779*, p. 192.
142. See Barr, Daniel P. (2006), "Epilogue: Modern Military Tradition," *Unconquered: The Iroquois League at War in Colonial America*, Praeger, Westport, CT, 165–68.
143. Although the historical marker indicates that the visit took place around 1820, the correct timing of his visit (from Louis Philippe's journals and those of his brothers who accompanied him) was during an extended exile in 1796–97. See Bell, B. (2005), *Schuyler County, New York: History and Families,* Schuyler County Historical Society, Turner Press, p. 19.
144. The best-known is Thomas Jefferson, who had children with his Negro house attendant slave Sally Hemings. See Foster, Eugene A. et al. (1998), Jefferson fathered slave's last child, *Nature*, 396, Nov. 5, p. 27.
145. From Turner, O. (1851), *History of the Pioneer Settlement of Phelps and Gorham's Purchase and Morris' Reserve; Embracing the Counties of Monroe, Ontario, Livingstone, Yates, Steuben, Most of Wayne and Allegany, and Parts of Orleans, Genesee, and Wyoming, to which is Added, A Supplement, or Extension of the Pioneer History of Monroe County*, William Alling, Rochester, NY, p. 99.
146. See the Cornell Web page, http://aip.cornell.edu/about/mission.cfm.

CHAPTER EIGHT: COMPETITION IS UNTENABLE

147. The first use of this term, often attributed to Charles Darwin, was by Malthus roughly sixty years before the publication of Darwin's *On the Origin of Species by Means of Natural Selection or the Preservation of Favoured Races in the Struggle for Life* (1859). The phrase appears in Malthus's "*Essay on Population . . .* " (1798) in the opening paragraphs of chapter 3, and "the struggle" is referred to throughout the essay. Charles Darwin read Malthus's essay in 1838, the sixth edition, and it formed a key insight for the theory of natural selection. See Desmond, Adrian J., and James R. Moore, (1991), *Darwin*, Michael Joseph, London, p. 264.
148. This quote is from chapter 8 of Malthus's first edition, 1798. See Appleman, P. (2004), ed., *An Essay on the Principle of Population: A Norton Critical Edition*, Norton, New York, p. 58.
149. Literary conventions of politeness at the time restricted Malthus from suggesting anything overtly sexual. He writes of moral restraint: "In modern Europe a much larger proportion of women pass a considerable part of their lives in the exercise of this virtue, than in past times and among uncivilized nations . . . implies principally a delay of the marriage union from prudential considerations . . . it may be considered in this light as the most powerful checks, which in modern Europe keep down the population to the level of subsistence." Malthus, Thomas R. (1817), *An Essay on the Principle of Population; or, a View of its Past and Present Effects on Human Happiness; with an Inquiry into Our Prospects Respecting the Future Removal or Mitigation of the Evils which it Occasions*, fifth ed., John Murray, London, vol. 2, p. 218.

150. See Ross, E. B. (2004), "The Malthus Factor," in *An Essay on the Principle of Population: A Norton Critical Edition*, edited by P. Appleman, Norton, New York, p. 238.
151. Nowhere is this better exemplified than in the ongoing debate about the airline industry. It's assumed that mergers lead to fewer competing airlines and will result in higher fare. But competition isn't the only way to reduce ticket prices. For instance, the government could regulate them. This isn't mentioned anymore in the press because the tacit understanding is that competition is the only game in town. But price attenuation through government subsidies works in agriculture (the dairy and corn industries for instance), so why not in the airlines? See *New York Times*, Opinion, "An Unwise Airline Merger," Nov. 14, 2013.
152. See Desmond, Adrian J., and James R., Moore, *Darwin*, p. 453.
153. Alfred Russel Wallace is considered the codiscoverer of natural selection. An observant naturalist who, like Darwin, traveled the world and spent years investigating the tropical rain forests, Wallace sent a letter to Darwin before the publication of *Origin of Species*. His letter outlined a view of natural selection that nearly matched the view Darwin came up with independently. Wallace, too, was inspired by what he read in Malthus's *Essay on Population*. Darwin accepted Wallace's ideas and agreed to present a paper jointly announcing their codiscovery of natural selection. After this Darwin stepped up his efforts to publish the more extensive treatment of the subject in *Origin of Species*. Hence the mechanism of evolution we recognize today is a blend of both their views. For evolution to occur, organisms must meet the basic requirements of their environment as well as outcompete other individuals for access to reproductive partners.
154. See Browne, E. J. (2003), *Charles Darwin, the Power of Place*, Knopf, New York, p. 187.
155. Darwin, Charles (1859), *On the Origin of Species by Means of Natural Selection . . .*, John Murray, London, p. 320.
156. See Raup, David M. (1991), *Extinction, Bad Genes or Bad Luck?,* Norton, New York, 1992.
157. Provine, W. B. (1989) "Evolution and the Foundation of Ethics," in *Science, Technology, and Social Progress*, edited by S. L. Goldman, Lehigh University Press, Bethlehem, PA, pp. 253–67.
158. Neo-Darwinism is distinguished from Darwinism and neo-Lamarckism. Darwin recognized that organisms often possessed trivial or apparently nonuseful traits that were faithfully passed from parent to offspring, and that such traits could not be explained by natural selection. This recognition became one of the most important hallmarks of Darwinism: Natural selection explained adaptation, but not all traits were adaptations. After he died, neo-Darwinism, or "ultra-selectionism" as Vernon Kellogg called it (see Kellogg, *Darwinism To-Day*, pp. 38–39), became more popular. The origin of this distinction between neo-Darwinism and Darwinism comes from August Weismann, who by proving the separation of germ cells and somatic cells established the grounds for a "new" Darwinian worldview. Neo-Darwinism finally freed Darwinism from the "taint of Lamarckism." Lamarckism is also called the "inheritance of acquired characteristics," the idea that life experiences of the parents could be incorporated into

the hereditary material. Weismann's view was that natural selection is the only mechanism driving evolution forward. This extreme view was never advocated by Darwin. Nor was the notion of absolute separation of the somatic and germ cells. Darwin believed in hereditary molecules (gemmules) that were strongly influenced by the somatic cells. This made Darwin's theory of heredity inherently Lamarckian. See reference in note 158, pp. 136 and 188, for further discussion.

159. Darwin, Charles (1868), *The Variation of Plants and Animals Under Domestication*, John Murray, London. My personal copy of this is the second edition of 1872. In this two-volume set, gemmules are introduced in volume 2, p. 372.
160. Kellogg, Vernon L. (1908), *Darwinism To-Day, a Discussion of Present-Day Scientific Criticism of the Darwinian Selection Theories, Together with a Brief Account of the Principal Other Proposed Auxiliary and Alternative Theories of Species-Forming*, Henry Holt and Co., New York. p. 199.
161. Kellogg, Vernon L., and Bell, Ruby Green Bell (1904), *Studies of Variation in Insects, Proceedings of the Washington Academy of Sciences*, vol. 6, 203–332.
162. Dobzhansky, Theodosius (1937), *Genetics and the Origin of Species*, Columbia University Press, New York.
163. Provine, W. B. (1989), "Progress in Evolution and Meaning in Life," in *Evolutionary Progress*, edited by Mathew H. Nitecki, University of Chicago Press, Chicago, p. 61.
164. Mayr, E., and W. B. Provine, eds. (1998), *The Evolutionary Synthesis, Perspectives on the Unification of Biology*, Harvard University Press, Cambridge; also Provine, W. B. (1989), "Progress in Evolution and Meaning in Life," in *Evolutionary Progress*, edited by Matthew H. Nitecki, University of Chicago Press, Chicago; and Gould, Stephen J. (2002), *The Structure of Evolutionary Theory*, Belknap Press, Cambridge, chap. 7.
165. Kimura, Motoo (1983), *The Neutral Theory of Molecular Evolution*, Cambridge University Press, Cambridge (UK). Motoo Kimura suggests that the main cause of evolution at the molecular level—that is, in the DNA—is not caused by natural selection but rather by random changes (mutations) that become fixed as selectively neutral variations.
166. Margulis, Lynn and Dorion Sagan (2002), *Acquiring Genomes: A Theory of the Origins of Species*, Basic Books, New York, p. 16.
167. Whitehead, Alfred North N. (1935), *Science and the Modern World*, Macmillan, New York, pp. 74–82.

CHAPTER NINE: KNOW THYSELF, DON'T LIE TO THYSELF

168. See http://www.cnn.com/2014/03/08/justice/florida-mother-minivan-ocean/ March 9, 2014; and http://www.nbcnews.com/storyline/minivan-mom-case/attempted-murder-charges-filed-against-mom-who-drove-kids-ocean-n47146, March 7, 2014.
169. Huang, Y. F., et al. (2014), Pre-existing brain states predict risky choices, *NeuroImage* 101, 466–72.
170. This number is one thousand times the number of stars in our galaxy, the Milky Way. See http://discovermagazine.com/2011/mar/10-numbers-the-nervous-system#.Uyn2AaldWmE.
171. See Edelman, Gerald M., and Giulio Tononi (2000), *A Universe of Consciousness: How Matter Becomes Imagination*, Basic Books, New York, especially

chs. 7 and 8; and LeDoux, Joseph E. (2002), *Synaptic Self: How Our Brains Become Who We Are*, Viking, New York, especially ch. 4.

172. This area is specifically called the PFC, or prefrontal cortex. It is one of five, six, or more (depending on the classification) regions of highly folded tissue that compose the human brain.
173. Roth, G., and U. Dicke (2005), Evolution of the brain and intelligence, *Trends in Neuroscience* 9, no. 5, 250.
174. Crick, Francis (1994) *The Astonishing Hypothesis*, Charles Scribner's Sons, New York.
175. Changeux, Jean P. (1985), *Neuronal Man, the Biology of Mind*, translated by Dr. Laurence Garey, Pantheon, New York, pp. 246–49.
176. Ridley, Jasper G. *The Tudor Age* (Constable & Robinson, 1998), pp. 69–90.
177. See Fodor, J. (2003), "Why Would Mother Nature Bother?," *London Review of Books* 25, no. 5, 17–18; and Dennett, D. (2003), "Letter in Response to Fodor," *London Review of Books* 25, no. 7.
178. Haladjian, Harry H, and Carlos Montemayor (2014), "On the evolution of conscious attention," *Psychonomic Bulletin and Review*, 22, no. 3, 595–613.
179. In 2010 I published my Ph.D. dissertation and retitled it *Evolution and Religion, Questioning the Beliefs of the World's Eminent Evolutionists*. Polypterus Press, Ithaca, NY.
180. Palumbi, S. (2001), "Humans as the World's Greatest Evolutionary Force," *Science* 293, no. 5536, 1786–90.
181. McClure, Nathan S., and Troy Day (2014), A theoretical examination of the relative importance of evolution management and drug development for managing resistance, *Proceedings of the Royal Society Publishing,* 281.
182. Lovelock, James (2009), *The Vanishing Face of Gaia,* Basic Books, New York.
183. The history of the American environmental movement goes back at least to the founding of the Sierra Club in 1892 and the Audubon Society in 1905. But in the 1960s the Environmental Defense Fund was involved in fresh water cleanup. Their study of the Mississippi River in 1974 led to the first legislation for the Safe Drinking Water Act of 1974.

CHAPTER TEN: EVOLUTION MANAGEMENT

184. http://www.botany.org/PlantScienceBulletin/psb-2004-50-4.php#Dodo.
185. http://content.govdelivery.com/accounts/USCDC/bulletins/bd32d6.
186. Althaus, Christian L. (2014), Estimating the Reproduction Number of Ebola Virus (EBOV) During the 2014 Outbreak in West Africa, *PLoS Current Outbreaks*, 2014 September 2, 6.
187. See http://www.bloomberg.com/news/2014-11-30/ebola-rages-in-sierra-leone-as-un-misses-goals-for-curbing-cases.html
188. This earthquake registered 9.0 on the Richter scale, the largest in Japan's history and the fifth-largest ever recorded in the world. It resulted from movement along a fault that moved the entire island of Honshu eight feet to the east. This island, Japan's largest and most populated, includes the cities of Tokyo, Nagoya, and Osaka. A forty-foot tsunami followed the earthquake and sent a wave of water nearly seven miles inland, inundating the city of Fukushima and its surroundings.

189. The age and sex of individuals in a population.
190. A taxon is a formal category recognized by taxonomists—such as the species, for instance. Taxonomists are professionals who name and study the biologically significant traits of organisms in order to classify distinct species, genera, families, orders, phyla, and so on.
191. See Gaboldon, T. (2010), Peroxisome diversity and evolution, *Philosophical Transactions of the Royal Society B* 365, 765; Schluter, A., et al. (2006), The evolutionary origin of peroxisomes: an ER-peroxisome connection, *Molecular Biology and Evolution* 23, no. 4, 838; and Cavalier-Smith, T., M. Brasier, and T. M. Embley, (2006), Introduction: How and when did microbes change the world? *Philosophical Transactions of the Royal Society* B 361, 845; and de Duve, Christian (2002), *Life Evolving, Molecules, Mind, and Meaning*, Oxford University Press, New York, pp. 145–48.
192. Intergovernmental Panel on Climate Change, www.IPCC.ch.
193. Klein, Naomi (2011), "Capitalism vs. the Climate," *The Nation*, Nov. 28, 2011.
194. See Royer, Dana L. (2006), CO_2-forced climate thresholds during the Phanerozoic, *Geochemica et Cosmichemica Acta* 70, 5665.
195. See National Climatic Data Center of the National Oceanic and Atmospheric Administration; www.ncdc.noaa.gov/sotc/global; Gillis, J. (2014), "The Flood Next Time," *New York Times*, Jan. 13, Environment Section, and Freedman, A. (2013), "Making sense of the Moore Tornado in a climate context, climatecentral.org, May 21; Shepherd, A., and Wingham, D. (2007), Recent sea-level contributions of the Antarctic and Greenland ice sheets, *Science* 315, no. 5818, 1529.
196. A recent study shows that some rugosans in the Silurian (443–418 mya), were tiny and endosymbiotically embedded into the skeletal tissue of stromatoporoids. Stromatoporoid reefs show extensive evidence of endosymbiosis with other species as well, such as *Cornulites sp*., an encrusting and burrowing wormlike animal. Other reef dwellers from this period show further evidence of endosymbiosis that does not involve stromatoporoids, Lingulid brachiopods (superficially resembling clams, but much different in their soft anatomy, with a muscular "foot" used for burrowing) inside of tabulate corals, for instance. Hence reef organisms have a long history of endosymbiosis that continues to this day, even though the species that form them through time have been different. See Vinn, O., and M. A. Motus (2014), Endobiotic Rugosan Symbionts in Stromatoporoids from the Sheinwoodian (Silurian) of Baltica, *PLoS One*, 25, 9, no. 2.
197. see Veron, John (2008), Mass extinctions and ocean acidification: Biological constraints on geological dilemmas, *Coral Reefs* 27, no. 3, 459–72.
198. http://earthobservatory.nasa.gov/Features/Coral/.
199. http://earthobservatory.nasa.gov/Features/Coral/coral2.php.
200. In addition to being the title of a popular book by Elizabeth Kolbert, and a Web site devoted to the "current biodiversity crisis," The Sixth Extinction has been batted around paleontology circles since I was a student. In addition to the "Big Five" extinctions in Earth history, the Ordovician (445 mya), the Devonian (350 mya), the Permo-Triassic (255 mya), the Triassic (200 mya), and the Cretaceous-Tertiary (65 mya), the Pleistocene (11,000 years ago) would be the sixth.
201. Galton, Francis (1892), *Hereditary Genius,* 2nd ed., Watts and Co., London, p. 1.
202. http://www.eugenicsarchive.org/html/eugenics/static/themes/39.html.

203. See Gould, Stephen J. (1996), *The Mismeasure of Man*, Norton, New York, p. 409.
204. Lewontin, Richard C. (2011), "It's Even Less in Your Genes," *New York Review of Books*, May 26. For a further guide to the issue of IQ it is worth reading the spirited exchange by Richard Lewontin, who summarizes the fallacy of the intelligence quotient beautifully, and his critics on the topic in "Is Intelligence for Real? An Exchange," *New York Review of Books*, October 22, 1981.
205. Microevolution is that branch of biology that deals with changes in populations from one generation to the next, and it is measurable in ecological time frames, on a scale of one to one hundred years, for instance. Macroevolution generally deals with higher-order changes in evolutionary lineages that occur over tens of thousands to millions of years and are seen in the fossil record. For instance, determining the number of different elephant species that have evolved over the last 20 million years falls in the realm of macroevolution; measuring the variation of wing patterns in a butterfly population over the course of ten years falls in the realm of microevolution.
206. For instance, the U.S. Department of Energy offers grants and awards to innovative proposals for twelve projects to test innovative concepts for the beneficial use of carbon dioxide. Some 1.4 billion dollars have been set aside for this program "from the American Recovery and Reinvestment Act (ARRA) for projects that will capture carbon dioxide from industrial sources." See http://energy.gov/fe/innovative-concepts-beneficial-reuse-carbon-dioxide-0.

REFERENCES

Abler, Thomas S. "Before the American Revolution." In *Cornplanter: Chief Warrior of the Allegany Senecas*, 15. Syracuse, NY: Syracuse University Press, 2007.

Al-Quadan, Tasneem, Christopher T. Price, and Yousef Abu Kwaik. "Exploitation of Evolutionarily Conserved Amoeba and Mammalian Processes by Legionella." *Trends in Microbiology* 20, no. 6 (2012): 299–306.

Althaus, Christian L. "Estimating the Reproduction Number of Ebola Virus (EBOV) During the 2014 Outbreak in West Africa." *PLOS Current Outbreaks*, September 2, 2014.

"An Unwise Airline Merger." *New York Times*. November 13, 2013. http://www.nytimes.com/2013/11/14/opinion/an-unwise-airline-merger.html?_r=0.

"Antibiotic Resistance Threats in the United States, 2013." Centers for Disease Control and Prevention. July 17, 2014. http://www.cdc.gov/drugresistance/threat-report-2013/.

Banchereau, Jacques. "The Long Arm of the Immune System." *Scientific American* 287, no. 5 (2002): 52–59.

Barr, Daniel P. "Epilogue: The Longhouse Endures." In *Unconquered: The Iroquois League at War in Colonial America*, 165–68. Westport, CT: Praeger, 2006.

Bell, Philip John Livingstone. "Viral Eukaryogenesis: Was the Ancestor of the Nucleus a Complex DNA Virus?" *Journal of Molecular Evolution* 53, no. 3 (2001): 251–56.

Bersaglieri, T., P. Sabeti, N. Patterson, T. Vanderploeg, S. Schaffner, J. Drake, M. Rhodes, D. Reich, and J. Hirschhorn. "Genetic Signatures of Strong Recent Positive Selection at the Lactase Gene." *American Journal of Human Genetics* 74, no. 6 (2004): 1111–20. http://www.ncbi.nlm.nih.gov/pmc/articles/PMC1182075/.

Boehm, Thomas. "Evolution of Vertebrate Immunity." *Current Biology* 22, no. 17 (2012): R722–32.

Botelho, Greg. "Sister: Mother who drove with children into ocean thought there were 'demons,' " CNN.com, March 9, 2014. Accessed February 27, 2015. http://www.cnn.com/2014/03/08/justice/florida-mother-minivan-ocean/.

Browne, E. J. "Eyes Among the Leaves." In *Charles Darwin: The Power of Place*, 187. New York: Alfred A. Knopf, 2002.

"Buck vs. Bell Trial." *EugenicsArchive.* http://www.eugenicsarchive.org/html/eugenics/static/themes/39.html.

Cavalier-Smith, T., M. Brasier, and T. M. Embley. "How and When Did Microbes Change the World?" *Philosophical Transactions of the Royal Society B: Biological Sciences* 361, no. 1470 (2006): 845–50.

Changeux, Jean. "To Learn Is to Eliminate." In *Neuronal Man: The Biology of Mind*, 2nd edition, 246–49. New York: Oxford University Press, 1997.

Cook, Frederick. "Journal of Officers." *Journals of the Military Expedition of Major General John Sullivan Against the Six Nations of Indians in 1779; With Records of Centennial Celebrations; Prepared Pursuant to Chapter 361, Laws of the State of New York, of 1885*, 8–11. Auburn, NY: Knapp, Peck & Thomson, Printers, 1887.

Cordaux, Richard, and Mark A. Batzer. "The Impact of Retrotransposons on Human Genome Evolution." *Nature Reviews Genetics* 10, no. 10 (2009): 691–703.

Crick, Francis. *The Astonishing Hypothesis: The Scientific Search for the Soul.* New York: Charles Scribner's Sons, 1995.

Cryan, John F., and Timothy G. Dinan. "Mind-altering Microorganisms: The Impact of the Gut Microbiota on Brain and Behaviour." *Nature Reviews Neuroscience* 13, no. 10 (2012): 701–12. http://www.ncbi.nlm.nih.gov/.

Darwin, Charles. *On the Origin of Species by Means of Natural Selection, or the Preservation of Favored Races in the Struggle for Life*. London: John Murray, 1860.

———. *The Variation of Animals and Plants under Domestication*. London: Murray, 1868.

David, Lawrence A., Corinne F. Maurice, Rachel N. Carmody, David B. Gootenberg, Julie E. Button, Benjamin E. Wolfe, Alisha V. Ling, A. Sloan Devlin, Yug Varma, Michael A. Fischbach, Sudha B. Biddinger, Rachel J. Dutton, and Peter J. Turnbaugh. "Diet Rapidly and Reproducibly Alters the Human Gut Microbiome." *Nature* 505 (2014): 559–63.

Dawkins, Richard. "Lampreys and Hagfish." In *The Ancestor's Tale: A Pilgrimage to the Dawn of Evolution*, 354. Boston: Houghton Mifflin, 2004.

Dennett, D. "Letter in Response to Fodor." *London Review of Books* 25, no. 7, April 2003.

Desmond, Adrian J., and James R. Moore. "Marriage and Malthusian Respectability." In *Darwin*, 264. London: Michael Joseph, 1991.

Dobzhansky, Theodosius. *Genetics and the Origin of Species*. 3rd ed. New York: Columbia University Press, 1951.

Doolittle, W. F., and O. Zhaxybayeva. "On the Origin of Prokaryotic Species." *Genome Research* 19, no. 5 (2009): 744–56.

Dufel, Susan E. "CBRNE—Plague." *Medscape.* December 20, 2013. Accessed February 27, 2015. http://emedicine.medscape.com/article/829233-overview#a0199.

Duve, Christian de. "The Mysterious Birth of Eukaryotes: A Possible Pathway." In *Life*

Evolving: Molecules, Mind, and Meaning, 145–48. New York: Oxford University Press, 2002.

Echenberg, Myron J. "Pestis Redux: The Initial Years of the Third Bubonic Plague Pandemic, 1894–1901." *Journal of World History* 13, no. 2 (2002): 429–49.

Edelman, Gerald M., and Giulio Tononi. " 'Selectionism' and 'Nonrepresentational Memory.' " In *A Universe of Consciousness: How Matter Becomes Imagination*, 7th edition, 79–101. New York: Basic Books, 2000.

Fedonkin, M. A., et al. *The Rise of Animals: Evolution and Diversification of the Kingdom Animalia*. Baltimore: Johns Hopkins University Press, 2007.

Feltman, Rachel. "The Gut's Microbiome Changes Rapidly with Diet." *Scientific American*, 2013.

Fischer, David Hackett. *Champlain's Dream: The European Founding of North America*. New York: Simon & Schuster, 2008.

Fischer, Joseph R. *A Well-Executed Failure: The Sullivan Campaign against the Iroquois, July–September 1779*. Columbia: University of South Carolina Press, 1997.

Fodor, J. "Why Would Mother Nature Bother?" *London Review of Books* 25, no. 5 (2003): 17–18.

Freedman, Andrew. "Making Sense of the Moore Tornado in a Climate Context | Climate Central." *Climate Central*. May 21, 2013. http://www.climatecentral.org/news/making-sense-of-the-moore-tornado-in-a-climate-context-16021.

Gabaldon, T. "Peroxisome Diversity and Evolution." *Philosophical Transactions of the Royal Society B: Biological Sciences* 365, no. 1541 (2010): 765–73.

Galton, Francis. *Hereditary Genius: An Inquiry into Its Laws and Consequences*. Cleveland, OH: Meridian Books, 1962.

Gause, G. F. "Experimental Studies on the Struggle for Existence: Mixed Populations of Two Species of Yeast." *Journal of Experimental Biology* 9, no. 4 (1932): 389–402.

Gillis, Justin. "The Flood Next Time." *New York Times*, January 13, 2014. http://www.nytimes.com/2014/01/14/science/earth/grappling-with-sea-level-rise-sooner-not-later.html?_r=0.

"Global Analysis—January 2015." *Global Analysis*. Accessed March 2, 2015. http://www.ncdc.noaa.gov/sotc/global.

Gould, Stephen Jay. *Ontogeny and Phylogeny*. Cambridge: Belknap Press of Harvard University Press, 1977.

———. *The Structure of Evolutionary Theory*. Cambridge: Belknap Press of Harvard University Press, 2002.

———. "Three Centuries' Perspectives on Race and Racism." In *The Mismeasure of Man*, revised and expanded edition, 409. New York: Norton, 1996.

Graffin, Greg. *Evolution and Religion: Questioning the Beliefs of the World's Eminent Evolutionists*. Ithaca, NY: Polypterus Press, 2010.

———, and Steve Olson. *Anarchy Evolution: Faith, Science, and Bad Religion in a World Without God*. New York: Harper Perennial, 2010.

Greer, Allan. "The Hurons Annihilated." In *The Jesuit Relations: Natives and Missionaries in Seventeenth-Century North America*. Boston: Bedford/St. Martin's, 2000.

Gupta, Radhey S. "Protein Phylogenies and Signature Sequences: A Reappraisal of Evolutionary Relationships among Archaebacteria, Eubacteria, and Eukaryotes." *Microbiology and Molecular Biology Reviews* 62, no. 4 (1998): 1435–91.

———, and G. Brian Golding. "The Origin of the Eukaryotic Cell." *Trends in Biochemical Sciences* 21, no. 5 (1996): 166–71.

Haladjian, Harry H., and Carlos Montemayor. "On the Evolution of Conscious Attention." *Psychonomic Bulletin and Review* 22, 3, 2014.

Harlow, Mary Laurence, and Ray Laurence. "Old Age in Ancient Rome: Mary Harlow and Ray Laurence Look at What It Meant to Become a Senior Citizen in Ancient Rome, and How This Early Model Has a Bearing on Our Attitudes Towards Ageing Today." *History Today* 53, 4, April 1, 2003.

"Highlights: Emerging Infectious Diseases." Centers for Disease Control and Prevention (CDC). June 6, 2014. http://content.govdelivery.com/accounts/USCDC/bulletins/bd32d6.

Hunt, George T. "The Hurons and Their Neighbors." In *The Wars of the Iroquois: A Study in Intertribal Trade Relations*, 3rd edition, 38–53. Madison: University of Wisconsin Press, 1960.

Huxley, Thomas Henry. "Page Evolution and Ethics: Prolegomena (1894)." In *Evolution and Ethics and Other Essays*, 8. New York: D. Appleton and Company, 1896.

"Innovative Concepts for Beneficial Reuse of Carbon Dioxide." Energy.gov: Office of Fossil Energy. http://energy.gov/fe/innovative-concepts-beneficial-reuse-carbon-dioxide-0.

"IPCC—Intergovernmental Panel on Climate Change." IPCC—Intergovernmental Panel on Climate Change. http://ipcc.ch/.

Jordan, Kurt A. *The Seneca Restoration, 1715–1754: An Iroquois Local Political Economy*. Gainesville: University Press of Florida, 2008.

Kellogg, Vernon L. *Darwinism To-day a Discussion of Present-day Scientific Criticism of the Darwinian Selection Theories: Together with a Brief Account of the Principal Other Proposed Auxillary and Alternative Theories of Species-forming*. London: Henry Holt and Company, 1907.

———, and Ruby Green Bell Smith. *Studies of Variation in Insects*. Proceedings of the Washington Academy of Sciences, vol. 6, 203–332, 1904.

Kimura, Motoo. *The Neutral Theory of Molecular Evolution*. Cambridge, UK: Cambridge University Press, 1983.

Klein, Naomi. "Capitalism vs. the Climate." *The Nation*, November 28, 2011.

Knoll, Andrew H. "The Cyanobacteria, Life's Microbial Heroes." In *Life on a Young Planet: The First Three Billion Years of Evolution on Earth*, 5th edition, 102. Princeton, NJ: Princeton University Press, 2005.

Leclerc, Michel, Nicolas Kresdorn, and Björn Rotter. "Evidence of Complement Genes in the Sea-star *Asterias Rubens*: Comparisons with the Sea Urchin." *Immunology Letters* 151, nos. 1–2 (2013): 68–70.

LeDoux, Joseph E. "Building the Brain." In *Synaptic Self: How Our Brains Become Who We Are*, 2nd edition, 65–96. New York: Viking, 2002.

Lee, Y. K. "Looking into Our Intestines." In *Who Are We? Microbes, the Puppet Masters!*, 29. Hackensack, NJ: World Scientific, 2009.

Lehninger, Albert L., and David L. Nelson. "Carbohydrates and Glycobiology." In *Lehninger Principles of Biochemistry*, 4th edition, 252. New York: W.H. Freeman, 2005.

Lewontin, Richard C. "It's Even Less in Your Genes." *New York Review of Books*, May 26, 2011.

"Life Span of Human Cells Defined: Most Cells Are Younger Than the Individual."

Times Higher Education. September 2, 2005. http://www.timeshighereducation.co.uk/198208.article.

Litman, Gary W., Jonathan P. Rast, and Sebastian D. Fugmann. "The Origins of Vertebrate Adaptive Immunity." *Nature Reviews Immunology* 10, no. 8 (2010): 543–53.

Loewen, James W. "Introduction." In *Lies My Teacher Told Me: Everything Your American History Textbook Got Wrong*, 12th edition, 1–10. New York: New Press, 2007.

Lovelock, James. *The Vanishing Face of Gaia: A Final Warning*. New York: Basic Books, 2010.

Malthus, T. R. *An Essay on the Principle of Population: Influences on Malthus, Selections from Malthus' Work, Nineteenth-century Comment, Malthus in the Twenty-first Century*, 2nd ed. New York: Norton, 2004.

Malthus, Thomas Robert. "Of Increasing Wealth, as It Affects the Condition of the Poor." In *An Essay on the Principle of Population Or, a View of Its past and Present Effects on Human Happiness: With an Inquiry into Our Prospects Respecting the Future Removal or Mitigation on the Evils Which It Occasions*, 218. 5th edition London: Murray, 1817.

Margulis, Lynn. "Archaeal-eubacterial Mergers in the Origin of Eukarya: Phylogenetic Classification of Life." *Proceedings of the National Academy of Sciences* 93, no. 3 (1996): 1071–76.

———. "The Chimeric Eukaryote: Origin of the Nucleus from the Karyomastigont in Amitochondriate Protists." *Proceedings of the National Academy of Sciences* 97, no. 13 (2000): 6954–59.

———, and Dorion Sagan. "Darwinsim Not Neo Darwinsim." In *Acquiring Genomes a Theory of the Origins of Species*, 16. New York: Basic Books, 2002.

Mayr, Ernst, and W. B. Provine. *The Evolutionary Synthesis: Perspectives on the Unification of Biology*. Cambridge: Harvard University Press, 1980.

McClam, Erin. "Attempted Murder Charges Filed Against Mom Who Drove Kids into Ocean." NBC News. March 7, 2014. Accessed February 27, 2015. http://www.nbcnews.com/storyline/minivan-mom-case/attempted-murder-charges-filed-against-mom-who-drove-kids-ocean-n47146.

McClure, Nathan S., and Troy Day. "A Theoretical Examination of the Relative Importance of Evolution Management and Drug Development for Managing Resistance." *Royal Society Publishing* 281, no. 1797 (2014).

Monier, Adam, Jean-Michel Claverie, and Hiroyuki Ogata. "Taxonomic Distribution of Large DNA Viruses in the Sea." *Genome Biology* 9, no. 7 (2008): R106.

Mora, Camilo, Derek P. Tittensor, Sina Adl, Alastair G. B. Simpson, Boris Worm, and Georgina M. Mace. "How Many Species Are There on Earth and in the Ocean?" *PLoS Biology* 9, 8 (2013) 1–8: E1001127. http://journals.plos.org/plosbiology/article?id=10.1371/journal.pbio.1001127.

Murphy, Kenneth, and Paul Travers. *Janeway's Immunobiology*. 8th ed. New York: Garland Science, 2012.

Nitecki, Matthew H. *Evolutionary Progress*. Chicago: University of Chicago Press, 1988.

Norton, A. Tiffany. "The Iroquois Confederacy." In *History of Sullivan's Campaign against the Iroquois; Being a Full Account of That Epoch of the Revolution.*, 10. Lima, NY: A.T. Norton, 1879.

O'Malley, Nick. "War-weary America Loses Appetite for Battle." *Sydney Morning Herald*, April 4, 2014. http://www.smh.com.au/world/warweary-america-loses-appetite-for-battle-20140403-zqql2.html.

Orzack, Steven, Robert M. Martin, and Terry Tomkow. "Is Intelligence for Real? An Exchange." *New York Review of Books*, February 4, 1982.

Palumbi, S. R. "Humans as the World's Greatest Evolutionary Force." *Science* 293, no. 5536 (2001): 1786–90.

Parmenter, Jon. "On the Journey, 1534–1634." In *The Edge of the Woods: Iroquoia, 1534–1701*, 15–17. East Lansing: Michigan State University Press, 2010. See also Jon Parmenter (1999) "Isabel Montour: Cultural Broker on the Frontiers of New York and Pennsylvania," in *The Human Tradition in Colonial America,* ed. Ian K. Steele and Nancy L. Rhoden (Wilmington, DE).

Preston, David L. "The Tree of Peace Planted." In *The Texture of Contact European and Indian Settler Communities on the Frontiers of Iroquoia, 1667–1783.* Lincoln: University of Nebraska Press, 2009.

Provine, W. B. (1989) *Evolution and the foundation of ethics*; in *Science, Technology, and Social Progress*, Goldman, S. L. ed., Lehigh University Press, Bethlehem, PA.

Raup, David M. *Extinction: Bad Genes or Bad Luck?* New York: Norton, 1992.

"Recidivism." National Institute of Justice. June 17, 2014. http://www.nij.gov/topics/corrections/recidivism/Pages/welcome.aspx.

Richter, Daniel K. *Beyond the Covenant Chain: The Iroquois and Their Neighbors in Indian North America, 1600–1800.* Syracuse, NY: Syracuse University Press, 1987.

———. *The Ordeal of the Longhouse: The Peoples of the Iroquois League in the Era of European Colonization.* Chapel Hill, NC: Published for the Institute of Early American History and Culture, Williamsburg, Virginia, by the University of North Carolina Press, 1992.

Ricklefs, Robert E., and Gary L. Miller. "Predation." In *Ecology*, 450–78. 4th ed. New York: W. H. Freeman, 2000.

Ridley, Jasper Godwin. *The Tudor Age.* Woodstock, NY: Overlook Press, 1990.

Rohwer, Forest. "Global Phage Diversity." *Cell* 113, no. 2 (2003): 141.

Ross, Valerie. "Numbers: The Nervous System, From 268-MPH Signals to Trillions of Synapses," *Discover Magazine,* May 15, 2011, Accessed February 27, 2015. http://discovermagazine.com/2011/mar/10-numbers-the-nervous-system#.Uyn2AaldWmE.

Roth, G., and U. Dicke. "Evolution of the Brain and Intelligence." *Trends in Cognitive Science* 9, no. 5 (2005): 250–57.

Royer, Dana L. "CO_2-forced Climate Thresholds during the Phanerozoic."*Geochimica Et Cosmochimica Acta* 70, no. 23 (2006): 5665–75.

Ryan, Don. "Portland, Ore. Rejects Adding Fluoride to Drinking Water." *USA Today*, May 22, 2013. http://www.usatoday.com/story/news/nation/2013/05/22/portland-fluoride-water/2350329/.

Ryan, Frank. *Virolution: The Most Important Evolutionary Book Since Dawkins' Selfish Gene*. New York: HarperCollins, 2009.

Samia, N. I., K. L. Kausrud, H. Heesterbeek, V. Ageyev, M. Begon, K.-S. Chan, and N. C. Stenseth. "Dynamics of the Plague-wildlife-human System in Central

Asia Are Controlled by Two Epidemiological Thresholds." *Proceedings of the National Academy of Sciences* 108, no. 35 (2011): 14527–32.

Sanjuan, Rafael, Miguel R. Nebot, Nicola Chirico, Louis M. Mansky, and Robert Belshaw. "Viral Mutation Rates." *Journal of Virology* 84, no. 19 (2010): 9733–48.

Schoolcraft, Henry Rowe. "Memoranda." In *Notes on the Iroquois: Or, Contributions to American History, Antiquities, and General Ethnology.* Albany, NY: E. H. Pease & Co., 1847.

Schluter, A. "The Evolutionary Origin of Peroxisomes: An ER-Peroxisome Connection." *Molecular Biology and Evolution* 23, no. 4 (2006): 838–45.

Scott, G. R. "Elevated Performance: The Unique Physiology of Birds That Fly at High Altitudes." *Journal of Experimental Biology* 214 (2011): 2455–62.

Shepherd, A., and D. Wingham. "Recent Sea-Level Contributions of the Antarctic and Greenland Ice Sheets." *Science* 315, no. 5818 (2007): 1529–32.

Smith, L. Courtney, Lori A. Clow, and David P. Terwilliger. "The Ancestral Complement System in Sea Urchins." *Immunological Reviews* 180, 1 (2001): 16–34.

Sompayrac, Lauren. *How the Immune System Works.* 4th ed. Hoboken, NJ: Wiley-Blackwell, 2008.

Spadafora, Corrado. "A Reverse Transcriptase-Dependent Mechanism Plays Central Roles in Fundamental Biological Processes." *Systems Biology in Reproductive Medicine* 54, no. 1 (2008): 11–21.

Spencer, Herbert. "The Convergence of the Evidences." In *The Principles of Biology,* Vol. 2, 478. London: William and Norgate, 1864.

Sun, W., D. A. Six, C. M. Reynolds, H. S. Chung, C. R. H. Raetz, and R. Curtiss. "Pathogenicity of Yersinia pestis Synthesis of 1-Dephosphorylated Lipid A." *Infection and Immunity* 81, no. 4 (2013): 1172–85.

"TB Silent Killer: An Unforgettable Portrait of Lives Forever Changed by Tuberculosis." PBS. March 25, 2014. Accessed February 27, 2015. http://www.pbs.org/wgbh/pages/frontline/tb-silent-killer/.

Ting, C-N., M. P. Rosenberg, C. M. Snow, L. C. Samuelson, and M. H. Meisler. "Endogenous retroviral sequences are required for tissue-specific expression of a human salivary amylase gene." *Genes & Development* 6, no. 8 (1992): 1457–65.

Tonegawa, Susumu, Christine Brack, Nobumichi Hozumi, Gaston Matthyssens, and Rita Schuller. "Dynamics of Immunoglobulin Genes." *Immunological Reviews* 36, 1 (2006): 73–94.

Turner, O. *History of the Pioneer Settlement of Phelps and Gorham's Purchase, and Morris' Reserve: Embracing the Counties of Monroe, Ontario, Livingston, Yates, Steuben, Most of Wayne and Allegany, and Parts of Orleans, Genesee, and Wyoming.* Rochester, NY: William Alling, 1851.

Veron, John. "Mass Extinctions and Ocean Acidification: Biological Constraints on Geological Dilemmas." *Coral Reefs* 27, 3 (2008): 459–72.

Villarreal, Luis P., and Guenther Witzany. "Viruses Are Essential Agents within the Roots and Stem of the Tree of Life."*Journal of Theoretical Biology* 262, no. 4 (2010): 698–710.

Vinn, O., and M. A. Motus. "Endobiotic Rugosan Symbionts in Stromatoporoids from the Sheinwoodian (Sirlurian) of Baltica." *PLoS One* 9, no. 2, 1–7 (2014).

"Visitor Statistics." Newsroom of the Smithsonian. Accessed February 27, 2015. http://newsdesk.si.edu/about/stats.

Walter, Jens, and Ruth Ley. "The Human Gut Microbiome: Ecology and Recent Evolutionary Changes." *Annual Review of Microbiology* 65 (2011): 411–29.

"We the People, Not We the Corporations." Move to Amend. Accessed February 27, 2015. https://movetoamend.org/.

Weier, John. "Mapping the Decline of Coral Reefs: Feature Articles." NASA: Earth Observatory. March 12, 2001. http://earthobservatory.nasa.gov/Features/Coral/.

Weiss, R. A., and J. P. Stoye. "Our Viral Inheritance." *Science* 340, no. 6134 (2013): 820–21.

Wendel, C. H. *The Allis-Chalmers Story.* Iola, WI: Krause Publications, 1988.

White, Matthew. *Atrocities: The 100 Deadliest Episodes in Human History.* New York: Norton, 2012.

Whitehead, Alfred North. "The Eighteenth Century." In *Science and the Modern World, 1925*, 83–109. New York: Macmillan, 1935.

"The Widespread Misconception That the Tambalacoque or Calvaria Tree Absolutely Required the Dodo Bird for Its Seeds to Germinate." *Botanical Society of America* 50, no. 4 (2004). http://www.botany.org/PlantScienceBulletin/psb-.

INDEX